Deborah K. O'Brien

About the Authors

DICK MORRIS served as Bill Clinton's political consultant for twenty years. A Fox News political analyst, he is the author of six *New York Times* bestsellers (all with Eileen McGann), including *Outrage,* and one *Washington Post* bestseller.

EILEEN MCGANN, an attorney and consultant, is CEO of Vote.com and Legalvote.com. She works with Dick on campaigns around the world, specializing in using the Internet to win elections. They live in Florida.

DISCARD

OUTRAGE

BY DICK MORRIS AND EILEEN MCGANN

Condi vs. Hillary

Because He Could

Rewriting History

BY DICK MORRIS

Off with Their Heads

Power Plays

Vote.com

The New Prince

Behind the Oval Office

Bum Rap on America's Cities

OUTRAGE

How Illegal Immigration, the United Nations,
Congressional Ripoffs, Student Loan Overcharges,
Tobacco Companies, Trade Protection, and Drug Companies
Are Ripping Us Off . . . and What to Do About It

DICK MORRIS
& EILEEN McGANN

HARPER

NEW YORK • LONDON • TORONTO • SYDNEY

SPS00230817
EISENDERSE ACADEMY HS LIBRARY
105 TROY STREET
SENECA FALLS, NY 13148

HARPER

A hardcover edition of this book was published in 2007 by HarperCollins Publishers.

OUTRAGE. Copyright © 2007 by Eileen McGann. All rights reserved. Printed in the United States of America. No part of this book may be used or reproduced in any manner whatsoever without written permission except in the case of brief quotations embodied in critical articles and reviews. For information address HarperCollins Publishers, 10 East 53rd Street, New York, NY 10022.

HarperCollins books may be purchased for educational, business, or sales promotional use. For information please write: Special Markets Department, HarperCollins Publishers, 10 East 53rd Street, New York, NY 10022.

FIRST HARPER PAPERBACK EDITION PUBLISHED 2008.

Designed by Kris Tobiassen

Library of Congress Cataloging-in-Publication Data is available upon request.

ISBN 978-0-06-137393-0

08 09 10 11 12 DIX/RRD 10 9 8 7 6 5 4 3 2 1

To Eugene J. Morris
—96 and still going—

CONTENTS

OUTRAGE

INTRODUCTION

**OUTRAGE [*n*]: an act that violates accepted standards of be-
havior; the anger and resentment aroused by injury or insult.**

—MERRIAM-WEBSTER'S NEW COLLEGIATE DICTIONARY,
11th Edition

To paraphrase Billy Joel, Americans are in an outraged state of mind.

We're angry. We're fed up with the way most institutions in our society operate, and we're appalled at the pervasive culture of corruption that's becoming more and more evident in all levels of government, in major corporations, and even in humanitarian organizations.

We're offended by the unmistakable bias of the mainstream media. Even more, we're sick of the way Washington insiders spin every political misstep for public consumption, while they scratch each other's backs to maintain the status quo.

We're tired of watching special interests use their money and power to distort the legislative process and buy elections. Finally, we're infuriated to see that such behavior has become so de rigueur that we're no longer shocked by it. We actually expect it!

All of this needs to change. Immediately.

Wherever we look, in every sector of our economy, at every level of government, and throughout the world of politics, it's obvious that the prevailing policies are deliberately designed to benefit an elite few at the expense of the rest of us. This systematic corruption has been going on for too long. It is time to turn our democracy back into what it was intended to be: no longer a government of the pampered congressmen, paid for by the lobbyists who pervert the process for the benefit of greedy special interests, but a government truly of the people, by the people, and for the people.

It's time for a transformation—a change we have the power to accomplish.

We've already seen the beginnings of such a change. That's why American voters sent a clear message in the 2006 midterm elections. Turning Congress on its head, we defeated thirty incumbent Republican members of the House of Representatives and six GOP senators, ending the ten-year Republican majority in both houses. Furious about the never-ending ethical scandals, the flagrant self-dealing, and the unproductive "do-nothing" Congress, voters desperately sought a clean slate—and they got it.

Unfortunately, new faces don't necessarily mean new rules. If we don't follow this shake-up with serious institutional reforms, we won't have much of a revolution. Shifting the players alone won't be enough to transform the system. And it's the system that needs a major overhaul. After all, the Democrats are hardly morally superior to the Republicans. They've had their own scandals, and they're equally beholden to their own particular special interests. The American people didn't choose them because of some positive belief in their message. We chose them because we were tired of the same old gang of thieves.

No, this recent election wasn't a mandate for the Democratic Party. It was a mandate for correction and reform. As they often have before, the American people voted for a slate of nonincumbents, and in this case that meant non-Republicans. It was definitely time for a change. If history is any guide, however, once the outsiders take power, they tend to forget about the very reforms that they promised. So now it's up to us, the voters, to keep the pressure on—to force both Democrats and Republicans to clean up the Congress and reform our business community and public organizations. Because we know they're not working. They're certainly not being held accountable. And unless we start to hold their feet to the fire by pushing the enactment of some serious ethical reforms, the 2006 election will go down in history as a sophisticated game of political musical chairs.

There are some encouraging opening signs. House Speaker Nancy Pelosi has forced through a series of reforms that curb some—though not most and certainly not all—of the abuses in her chamber. She has success-fully induced her colleagues to ban travel and gifts from lobbyists . . . though congressional wives or husbands can still double-dip by going on their spouses' campaign payrolls or working for lobbying firms. She's re-quired congressmen to identify the wasteful earmarked appropriations they cram into spending bills . . . though she hasn't curbed the abuse by giving the president the power to veto or ignore the bloated spending amendments. Still, she has done something about the culture of congres-sional corruption, which is a lot more than her predecessor—Dennis Hastert—ever did.

The Senate reforms have led to serious changes, too. All corporate—and lobbyist—sponsored travel is now prohibited, as is the use of corporate jets by senators under any circumstances. In the House, new rules simply re-quire members to reimburse the corporate sponsor for the actual cost of the flights, not just the first-class airfare.

These new rules are a step in the right direction, but we deserve much more. Before Congress can restore confidence in the integrity of its legisla-tive actions, it must earn the trust of the American people.

That's no small job.

We're skeptical. We've learned too often that corporations are cooking their books, humanitarian agencies are on the take, and congressmen are taking bribes and secretly increasing their net worth. We've felt swindled too often after watching our investments dwindle while corporate execu-tives go around stealing everything that isn't nailed down. We've grown re-sentful that they get away with it. And we want it to stop.

We're revolted by the endless daily barrage of news stories chronicling fraud and abuse among our leaders. We cringed to learn that the United Na-tions looked the other way while Saddam Hussein stole almost $2 billion of humanitarian aid funds designated for emergency food and medicine for the Iraqi people in the UN's so-called Oil-for-Food Program. And we bris-tled to learn that the UN ignored Saddam's bribery of Russia and France to buy their support against the United States in the Security Council. Re-member France's then-ambassador to the UN, Dominique de Villepin, and the impassioned speech he made against intervention in Iraq? Turns out, he wasn't motivated by any fiery socialist idealism. No, his ulterior motive was calculated capitalist realism. France had become Iraq's largest trading partner, and de Villepin didn't want to jeopardize that cozy and lucrative relationship. Moreover, Saddam had given several French politicians—including Chirac's former interior minister, Charles Pasqua [1]—vouchers to sell Iraqi oil. Unfortunately for the French and Russians involved, Saddam also kept a ledger of these little favors—a document that later revealed the depths of these sneaky deals.

RUSSIA, FRANCE, SADDAM, AND THE UN:

HOW THE OIL-FOR-FOOD MONEY FLOWED TO CORRUPT POLITICIANS

What France Got:

- Vouchers for 11 million barrels of oil to former interior minister Charles Pasqua (each voucher for 10 million gallons could generate between $1 and $3.5 million in revenue) [2]
- Rights to develop a rich Iraqi oil field
- Payoffs to top French politicians, including diplomat Jean-Bernard Merimee (who holds the official title of "ambassador for life"), who took $156,000 in bribes to renovate his Moroccan vacation home while he served as a "Special Adviser" to UN secretary-general Kofi Annan [3]

What Russia Got:

- Vouchers for 55 million barrels of oil to the Russian foreign ministry
- Vouchers for 53 million barrels to Vladimir Zhirinovsky, former Russian presidential candidate and deputy speaker of the Duma (the Russian equivalent of the U.S. House of Representatives)
- Vouchers for 110 million barrels to the Russian Communist Party
- Vouchers to a former chief of staff to Vladimir Putin [4]
- Vouchers to the Russian Orthodox Church
- Rights to another Iraqi oil field

What the UN Got:

- $1.9 billion in cash from Iraqi oil sales
- 13 million barrels of oil to the head of the Oil-for-Food Program, Benon Sevan, later indicted on bribery charges[5]

What Kojo Annan, Son of the UN Secretary-General, Got:

- $2,500 per month for four years after he left the employment of a Geneva company that got an Oil-for-Food contract while he worked for them[6]

What Saddam Got:

- More than $12 billion in cash for his personal use

What the Iraqi People Got:

- SCREWED! After all the bribery, graft, and outright theft, all they got were the leftovers.

And that's not all. During the entire time the UN officials were supervising the Oil-for-Food program, they never even noticed when Saddam shook down another $1.8 billion from corporations and government officials who were bidding to provide the food and other necessities to his starving citizens. It's especially hard to understand why UN officials did nothing about their own employees who were on the take—or why Secretary-General Annan had no idea that his own son was riding the profitable bandwagon of surreptitious payoffs.

What an outrage!

During the past several years, we've witnessed the shocking bankruptcies of Enron and WorldCom. With respective assets of $64 billion and $104 billion at the time of the bankruptcy filings, the two companies had been among the largest and most successful corporations in the United States. Because of the greed and fraud that permeated the highest echelons of these international businesses, thousands of hardworking men and women lost their jobs, their pensions, and their hopes for the future. Yet so far only a handful of the guilty executives who bled those companies dry have been prosecuted and convicted for those crimes, which changed the lives of so many. Another outrage.

And it's not just public corporations that are the problem. Recently it's come to light that some of our government-regulated agencies have been just as crooked. Look at Fannie Mae, the federally chartered organization owned by private shareholders that purchases mortgages in order to increase the availability of mortgage funding for low- and middle-income housing. In 1998, Fannie Mae failed to report more than $200 million in expenses; from 2000 to 2003, it overstated its earnings by more than $12 billion. Why did they engage in this financial trickery? Because its top executives' compensation packages—that is, their excessive bonuses—were tied

THE AMAZING DEMOCRATIC GRAVY TRAIN AT FANNIE MAE

- Franklin Raines, Fannie Mae's CEO, collected $90 million in bonuses and salary
- Jamie Gorelick, its vice chair, got $25 million
- Jim Johnson, its former CEO, got $2.9 million
- Hundreds of liberal nonprofits were given $35 million a year in grants

Raines, Bill Clinton's former budget director, was forced to resign, but he kept his $90 million; now he ekes out a meager living on a $114,000 monthly pension, part of a total retirement package estimated at $25 million![7] According to the *Rocky Mountain News,* former Clinton assistant attorney general Jamie Gorelick got more than $25 million in compensation, including $15 million in bonuses.[8] Jim Johnson, a Democratic insider who served as Walter Mondale's 1984 campaign manager, received $966,000 in salary and a $1,932,000 bonus in 1998.[9]

According to federal regulators, however, if Fannie Mae had been keeping accurate accounting, it would have paid *no bonuses at all* that year.

Now, that's an outrage!

to the company's performance. No profits, no bonuses. So the only way the folks in charge of Fannie Mae could line their pockets with obscene bonuses was to make the bottom line look great.

Congress is in on the game, too. And it's not just one party that's guilty. In recent months, Democrats and Republicans alike have been caught in major corruption scandals. One member of Congress was actually caught hiding $90,000 in cash in his home freezer (giving new meaning to the term *frozen assets*) and was apparently taped by the FBI as he accepted a $100,000 bribe. Louisiana Democratic congressman William Jefferson allegedly demanded cash payments and other favors for himself and his family in exchange for using his congressional position to help advance an African business scheme. After finding the frozen dollars, the Justice Department executed a search warrant on his congressional office. Republican and Democratic House leaders alike were furious, castigating the FBI for daring to search a congressman's office and screaming about the separation of powers. Apparently, they view the House of Representatives as an asylum for criminal activity beyond the reach of the United States Department of Justice. Former House speaker Dennis Hastert and his successor, Nancy Pelosi, both righteously claimed that the documents were unconstitutionally seized from Jefferson's office and demanded their immediate return. Now, isn't that outrageous? Shouldn't our congressional leaders want to *investigate* alleged criminal activity in their midst, rather than closing ranks around a suspect?

This bunker mentality in defense of institutional prerogative is common to both parties, and it may grow as more members come under criminal investigation.

And there are plenty to choose from. Former House majority leader Tom Delay (R-TX) resigned after he was indicted on money laundering charges and implicated in the Jack Abramoff lobbying scandal, one of the most amazing examples of special-interest influence peddling in the history of our nation. Former Republican congressman Duke Cunningham (R-CA) recently went to prison for accepting bribes from government contractors. As he was sentenced, he boldly named other members of Congress he claimed had also taken bribes! Ohio Republican congressman Bob Ney resigned in November 2006 after pleading guilty to making false statements and accepting trips, meals, airline tickets, cash, and gaming chips in exchange for pushing legislation for Abramoff's clients. And Congressman Mark Foley (R-FL) also resigned after details exposed his relationships with male high school students serving as House pages.

Stay tuned . . .

WHO POLICES THE COPS?

The Escapades of Congressman Alan Mollohan— Top Democrat on the House Ethics Committee

- From 2000 to 2005, Mollohan's personal assets rose from less than $500,000 to more than $8 million—all on a salary of $162,500 per year!
- He steered $5 million in federal earmarked funds to a nonprofit organization
- Then he bought a 300-acre farm with the head of the nonprofit organization
- Asked about the relationship, he claimed to see "no conflict of interest" [10]

Currently under federal investigation, Congressman Alan Mollohan was much more creative than his House colleagues. Even as he was serving as the ranking member on the House Ethics Committee, Mollohan saw an amazing increase in his personal assets—from less than half a million dollars in 2000 to somewhere between $6.3 and $24.9 million in 2005. [11]

How did he achieve this amazing return while making only $162,500 in his day job? One explanation could be what's known as *earmarking*—the process by which legislators steer specific federal funding to their favorite projects. Congressman Mollohan has a great record of earmarking: in the past decade, more than $250 million has been directed to nonprofit organizations he created in his native state. In most cases, the local organizations are staffed by the congressman's friends and former employees. Mollohan's recent real estate investments include the purchase of a 300-acre farm, with partners whose companies received earmarked funds channeled by Mollohan. What are friends for?

When asked to defend the propriety of a new partnership with a person whose company had received almost $5 million in earmarked federal funds, however, Mollohan said it had never occurred to him that there might be a conflict of interest.

Obviously, the former ranking member of the Ethics Committee was appointed to the wrong committee. Now, that's definitely an outrage!

The recent ethics changes in the House are a step in the right direction, but only a step. There's more to be done. Lately, we've become increasingly infuriated by the transparent and unabashed way our elected officials promote the agendas of the special interests that pay for their campaigns—while virtually ignoring the needs of constituents who elected them, and who depend on them for a voice in Washington.

They've created an Imperial Congress in Washington, with donors and lobbyists as their fawning court.

THE IMPERIAL CONGRESS

It's no surprise that Washington is geared to accommodate the powerful, not the average citizen in need of protection. But these days you get the feeling that our leaders don't really even want to see or hear from their constituents—literally!

New Jersey's multimillionaire senator Frank Lautenberg recently lamented the problems caused by ordinary people daring to use the "Senators Only" elevator in the Capitol.

According to the *New York Times,* Senator Lautenberg said that "he had never seen the Capitol so crowded with unelected interlopers." [12] Life's tough when you're a senator, isn't it, Frank? Nodding "gravely," he told the newspaper, "I hesitate to say that it's a big problem." [13]

Imagine: American citizens might want to visit their Capitol and see how it works—or how it *doesn't* work, as the case may be.

Senator Lautenberg isn't alone. Other senators, like former Ku Klux Klanner Robert Byrd (D-WV) and oil-drilling champion Ted Stevens (R-AK), reportedly greet their unwanted citizen visitors with "hostile glares."[14] As the *Times* points out, senators who stoop to take the regular elevators and mix with the hoi polloi also have to push their own elevator buttons, instead of relying on the friendly elevator operators. And that's just too much to ask from a United States senator. Lautenberg claims that the reason for the segregated elevators is to make sure that senators can quickly get to the floor to vote. Given how little time they actually spend in session, special elevators are hardly a necessity; rather, they're simply a perk to insulate them from annoying citizens.

MESSAGE TO CONGRESS

These "unelected interlopers" are your constituents—the ones who pay your salary, the ones who actually decide whether you stay or go!

Many senators and congressmen prefer to spend their time with lobbyists instead of with pesky constituents who just don't seem to know their place. But when powerful and rich special interests talk, Congress listens. (They don't even mind if lobbyists take the elite elevators!)

One classic case study is the story of how Congress handled the issue of raising the minimum wage. Republican leaders in Congress recently refused to increase the $5.15 hourly federal minimum wage to $7.25 for the lowest paid employees in our workforce, unless the measure was combined with a significant liberalization of the estate tax—that is, directly tied to a parallel benefit for the wealthiest families in the country. (The House bill would have repealed the estate tax completely, whereas the Senate bill would have increased the exemption from the current $2 million to $5 million.) Does that tell you who's pulling the strings? Insisting that the minimum wage be considered with the estate tax repeal, former Senate majority leader Bill Frist insisted that the minimum wage be bundled in with the estate tax repeal: it was "all or nothing" on the "death tax," he said—quid pro quo.[15]

Fortunately, the House passed the higher minimum wage bill anyway (at this writing, its ultimate fate remains unclear). But the very fact that Congress would think of linking a repeal of the estate tax to a raise in the minimum wage symbolized all that was wrong with the power special interests hold in America today.

Consider this: a raise in the minimum wage would mean that 6.6 million workers[16] would immediately bring home more money in their paychecks. Why would the members of Congress consider rejecting this measure to help them? Perhaps because the majority of these workers are white women between the ages of 24 and 64, who have high school diplomas or less and who work part-time.[17] Contrary to conventional wisdom, these are not all teenagers—temporary members of the workforce—working at fast-food restaurants. More than a million of them are single parents. They need the extra money to take care of their children! And they don't make campaign contributions.

In contrast, a total repeal of the estate tax would ultimately benefit just 8,200 people annually—while adding a trillion dollars to the national debt in the next fifteen years.[18] Talk about special-interest legislation! Under its rule, megamillionaires and billionaires would have nothing to fear; their interests would be protected. And if they weren't—well, the minimum-wage workers weren't getting protected, either. It was that simple.

As it is, very few people are required to pay the estate tax. According to the IRS, only 1.17 percent of all estates in 2002 were liable for the tax.[19] To put it in perspective, the estates of 99.7 percent of all people who died in the United States in 2006 will *not* be subject to the estate tax.[20] Only one-fourth of one percent of estates will pay the tax. That's due in large part to the important changes in the estate tax over the past few years to protect most American families.

In fact, under current law, the first $2 million of an estate is completely exempt from federal estate taxes. And when the exemption reaches its statutory increase to $3.5 million in 2009, only an estimated 2,500 people will be required to pay it each year.[21]

So what's the big deal? The big deal is that a few of the wealthiest families in the United States don't want to pay *any* taxes at all on their huge estates. So they've made sure that when they whistle, Congress dances.

REPEALING THE ESTATE TAX:

CHARITY FOR 8,200 PEOPLE

Total Cost? $1 Trillion

Who Benefits?

	Savings?
Charles and David Koch (Koch Industries)	$4.7 billion
The Mars Family (Mars Incorporated)	$3.9 billion
The Walton Family (Wal-Mart)	$1 billion plus
The Cox Family (Cox Communications)	$4.8 billion

Source: Conor Kenny, Taylor Lincoln, Chuck Collins, and Lee Farris, "Spending Millions to Save Billions: The Campaign of the Super Wealthy to Kill the Estate Tax," Public Citizen, April 2006, available at www.citizen.org/documents/EstateTaxFinal.pdf

Over the last ten years, a small number of mega-rich families have organized intensive lobbying efforts on the estate tax issue. According to a report by Public Citizen, a watchdog group, eighteen families have spent more than $200 million on this effort in the last ten years.[22] Public Citizen identifies the families who own Wal-Mart, the Mars candy conglomerate, the Gallo wine company, Nordstrom department stores, Koch Industries, Campbell Soup, Cox Communications, and others as the "stealth" lobbyists fueling the estate tax repeal. According to Joan Claybrook, president of Public Citizen:

The long-running, secretive campaign of eighteen extraordinarily wealthy families worth $185.5 billion has relied on deception to bamboozle the public and Congress not only about who must pay the estate tax, but about how repealing it will affect the country. . . . The families have hidden behind trade associations and lobbyists to make their pitch. They have paid for misleading, fear-inducing ads. They have poured tens of millions of dollars into their efforts, essentially buying what they want in Washington since 1998. They have pumped nearly $28 million into political campaigns to ensure doors open for them and their lobbyists, and collectively, they and their businesses have spent $27 million lobbying Washington officials on a variety of issues, including estate tax repeal.[23]

According to Public Citizen's estimates, the estate tax on these families adds up to an extraordinary amount of cash. Based on their estimated net worth of $24 billion, for instance, Charles and David Koch should have an estimated estate tax liability of more than $4.7 billion; the members of the Mars family would be close behind at $3.9 billion on their fortune of over $30 billion. The assessment on the heirs to Sam Walton's fortune would range from about $1 billion to $6 billion on their estimated assets of $83.7 billion. And the Cox Communications family, with assets of $24 billion, would have an estimated estate tax of $4.8 billion.[24]

With such amounts at stake, these families have taken no chances. Since 1999, they have contributed a total of $27,702,463 to political campaigns and organizations. The largest contributors in this group are the Wal-Mart Waltons, who gave more than $10 million during that time. Close behind is the DeVos family of Amway fame, which gave more than $6 million during that time. The Koch brothers gave more than $4 million.[25] Together, these families and their related entities spent another $27 million on direct lobbying on the estate tax and other issues.

And did it pay off? You be the judge. Take the example of Blanche Lincoln (D-AR), one of only three Democrats to support the estate tax changes: She received more than $100,000 from the super-wealthy families and their PACs since 1999. Or take Senator Max Baucus (D-MN): after the estate-tax repeal proponents mounted an aggressive campaign to unseat him, he jumped ship and joined the Republicans in supporting repeal.

While Congress was taking care of the richest people in the nation—at the expense of the very poorest people—they, of course, took especially good care of themselves. During the nine years in which they declined to raise the minimum wage, Congress increased its own pay *seven times,* for a total raise of $31,000. That's an outrage.

Is it any wonder, then, that the American people severely disapprove of the job Congress has done? Our lawmakers hardly work at all—and when they do, they're usually scrambling to please their major donors. And, of course, they make sure that their own interests are protected—and expanded.

In every major survey conducted in July and August 2006, voters gave Congress highly negative job approval ratings. In fact, on average only one-quarter of the respondents expressed approval.

AMERICANS DISAPPROVE OF CONGRESS. BIG TIME.

When asked, "Do you approve or disapprove of the job that Congress is doing?" voters responded as follows:

Poll	Approve [%]	Disapprove [%]	Not Sure [%]
Fox News (Aug. '06)	24	58	17
Fox News (July '06)	25	61	14
AP/Ipsos (Aug. '06)	29	69	1
AP/Ipsos (July '06)	27	68	1
New York Times (July '06)	28	58	0
Wall Street Journal	25	60	13
Diageo Hotline	31	58	11

The lowest negative rating in all of the polls was 58 percent; at its worst, it ranged as high as 69 percent. It was an astonishing vote of no confidence in our national legislators—and a harbinger of the election results a few months later.

This nationwide disdain for Congress is relatively new. According to the Gallup News Service, the current low ratings for Congress set "a near-record low for the institution. Gallup's trend for this question, which started in 1974, shows lower approval scores on only three other occasions: October 1994 (21%), March 1992 (18%), and June 1979 (19%)."[26]

What's behind these low ratings? Where to begin? Let's see: the absence of a meaningful legislative agenda; the disclosure of rampant institutional corruption; the lack of any serious competition against incumbents who had redistricted themselves into permanent seats; the emergence of an

"Imperial Congress." Seeing itself as a putative royal body, Congress has become populated by privileged members who see the extensive perks of office—free meals, trips, gifts for themselves and their families—as entitlements.

It has become a body of the rich—and of members of political dynasties. It stands to reason that members of Congress, especially in the Senate, may have a view of repealing the estate tax that is compatible with their superrich donors, since so many senators are millionaires themselves. *More than half of that body are now millionaires.* Another 11 percent are either wives or children of former senators, governors, or presidents; it has become a home for privileged political legacies.

Some members of Congress who support the minimum wage increase have suggested that congressional pay raises be tied to minimum wage increases. If there's no increase in the minimum wage, there could be no increase in congressional salaries. Any increase in one would trigger the same percentage increase in the other. That's a good idea. But we've got a better one: since members of Congress find the minimum wage so acceptable, how about paying it to them?

This raises the issue of congressional schedules. In 2006, the House worked only 103 days in official session. That's less than two days a week. Nice work, if you can get it. But, given their average workweek of two days or less, their minimum wage salary would be well below the poverty level.

That's about what most of them deserve.

Again, House Speaker Nancy Pelosi came in promising reform, pledging to keep the House in session five days each week (presumably beginning late Monday and ending early Friday so they can scurry home for the weekend). But Pelosi's promise was an empty one: The House is still meeting only two to three days each week. The Senate's schedule is only slightly better.

And Pelosi began her tenure by breaking her own rule to give House members Monday, January 8, off. Why? So they could watch the collegiate national championship football game!

So are we on the path to reform, or the same old road to ruin?

As far as we're concerned, the jury is out. So in the pages that follow we're offering a chronicle of the outrages we've seen in government and business. And we'll do more than just describe them. We've come up with a series of concrete proposals to stop the outrages, to turn our country and our society around, and to help us all control our own future. We hope it'll help explain why we feel victimized and ripped off—and what we can do to change that old feeling.

IMMIGRATION:
The Wide Open Door

For the average American, the issue of immigration is synonymous with the U.S.-Mexican border and the difficulty of curtailing illegal crossings over our vast southern boundary of rivers and deserts. Images come immediately to mind: droves of men, packed inside trucks, seeking to enter our country to find work and new opportunities. As a nation, we have a lot invested in protecting our perimeter from uninvited—and unwanted—visitors. Volunteer "Minutemen" patrol the border, acting as the eyes and ears of the Border Patrol. The Customs and Border Protection Division of Homeland Security deploys 11,300 guards on the frontier, and President Bush has recently ordered another 6,000 National Guard troops to join them.

This story of outwitting the illegal immigrants who secretly cross over from Mexico has received all of the attention and generated most of the heat in the current debate. But it is only one half of the problem of illegal immigration—and not the most important half. The biggest problem we face is caused by those who openly enter our country for a limited period of time—with the express permission of our government—and then refuse to leave.

On 9/11, we learned just how serious this problem is. It hasn't gotten any better.

It's time for America to adopt an aggressive program to identify and deport people who have deliberately and illegally overstayed their visas. Unless we do so, our efforts to secure our country will fail.

Consider this: of the 11.5 million illegal immigrants who lived in the United States in 2005, half did not cross the border with guards nipping at their heels. Instead, they came here *legally*, arriving by plane, train, car, or boat, with U.S. government entry visas in hand. Depending on their country of origin and their purpose in coming to the United States, these visitors were allowed to stay for a specific period of time. When their visas expired, however, they did not go home. Such visa overstays account for an estimated 50 percent of the illegal immigrants now living in the United States.[1]

Remember, the 9/11 terrorists were in this immigrant category—people who entered our country legally, with legitimate visas. Seven of the terrorists had illegally overstayed their visa time limits.

Despite the well-documented problems with these temporary-visitors-turned-permanent-immigrants, critics of our system generally ignore them and continue their preoccupation with the permeable Mexican border alone. The vocal condemnation of illegal aliens by the Immigration and Customs Enforcement (ICE), members of Congress, media talk shows, and the White House generally makes no distinction between those who entered with visas and those who sneaked across our border.

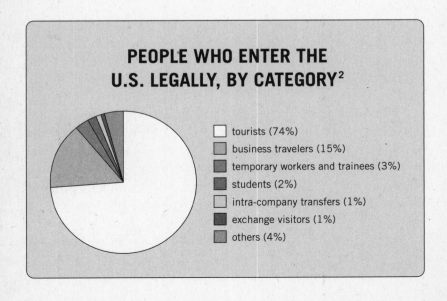

PEOPLE WHO ENTER THE U.S. LEGALLY, BY CATEGORY[2]

- tourists (74%)
- business travelers (15%)
- temporary workers and trainees (3%)
- students (2%)
- intra-company transfers (1%)
- exchange visitors (1%)
- others (4%)

When it comes to immigrants, in other words, America is like some-body who puts bars on his windows, securely bolts his back and side doors, and then leaves the front door unlocked and wide open.

We spend enormous resources policing our borders, but we do little or nothing to make sure that visitors whose visas have expired, do, in fact, leave our country.

Make no mistake about it: the door *is* wide open! We legally admit 33 million[3] people a year into the country, most of them with visas. This is up from 9.5 million in 1985. Sometimes they use forged or fraudulent docu-ments to gain entry. We try to detect them, but we're not always successful. Frequently their stated reasons for coming here have not been fully verified. It's our policy to check up on them, but sometimes we can't reach their ref-erences. But at least we try.

There's one critical thing that we don't do at all, however. We don't kick out those people who remain here after their legally issued visas expire. Once they cross the threshold, we don't pay much attention to them.

So about four to five million people who have no right to be here—who were granted permission to stay in the United States for only a limited time—are still here, and unlikely ever to leave. And it gets worse. Not only will many of them overstay their welcome, but lots of them will formally ask the government to change their status and let them remain, even though they're here illegally. And guess what? The government actually says *yes* to hundreds of thousands of them every year! Two of the nineteen 9/11 hi-jackers were actually granted status changes. One of these, terrorist leader Mohamed Atta, had long overstayed his visa.

According to a study by immigration expert Jessica Vaughan, "in 2001, more than 60 percent of all those who obtained permanent residency (653,259 out of 1,064,318) did so not by obtaining an immigrant visa, but through an 'adjustment of status,' which means that they were already pre-sent in the United States, sometimes legally, sometimes not."[4]

For the 650,000 people every year who change their status from tempo-rary to permanent, the strategy is much easier than the alternatives. Why wait on line for the immigration quotas to become available? Why hassle with crossing illegally and at great risk? Instead, they just come here as tourists or on work visas and stick around, knowing that the government will probably let them stay.

ICE, now under the Department of Homeland Security, works hard to protect our borders, to check the identities of those who arrive on our shores, and to stop those who might turn into permanent immigrants from coming to the United States. But it does almost nothing to be sure that, once people enter on visas, they actually leave when they're supposed to.

This should not be a difficult task. In fact, it would be very simple to confirm the names of those who leave as required, and thus of those who have blatantly ignored their visa requirements and remained in the States illegally. For example, ICE could initiate a system of checking the identities of noncitizens as they leave the United States, and then simply subtract their names from the list of those who had previously arrived on temporary visas. The remaining names would be a master list of those who had no right to be here. We could accomplish this by checking fingerprints as visitors leave the country, just as we now do on the way *in*. Under current US-VISIT policies, most immigrants have both index fingers fingerprinted as they arrive. In addition, all visitors have a digital photo taken, which is then compared with and matched to their passports and other travel documents. If that program were expanded to verify their identities on the way *out*, that would be another way to monitor expired visas effectively. But ICE doesn't have any nationwide system or policy of that sort in place.

Last year, very few people leaving this country were fingerprinted as they left. Customs agents have discretion to fingerprint and photograph green-card holders at land ports, but technical and policy hurdles have limited the number of people fingerprinted to roughly four million.

Crystal Williams, the deputy director of the American Immigration Lawyers Association, has lamented the absence of any effective process for tracking who is leaving the country and who is not. "They've only got one-half of the US-VISIT program working at all. You would have thought they would be concentrating on a viable exit solution," Williams said.[5]

Why aren't they? Our guess is that the government doesn't *want* to know—because it would find itself unable to deal with the huge backlog of people who are staying here illegally.

We've seen this before. When Dick worked for President Clinton, he proposed that illegal immigrants be prohibited from receiving driver's licenses and that traffic cops be equipped with computers that could tap into the INS database. As he explained to the president, this would make the average traffic police officer an enforcer of the immigration law. After all, the most frequent contact regular people have with the police is on the road. The system he was proposing would have allowed traffic officers to determine whether a traffic violator was here legally—or not.

Congress has since passed this measure into law, but, at the time President Clinton—and, particularly then first lady Hillary Clinton—as well as George Stephanopoulos, all opposed the idea, concerned that it would lead to racial profiling . . . and alienate the Clintons' Hispanic political base.

But the Clintons weren't the only ones opposed to the idea. Harold Ickes, then the White House deputy chief of staff, checked out the idea with the INS—and discovered that they opposed it, too. And their reason was not political at all. Rather, they feared the idea because they already had an unmanageable backlog of deportation cases on their plate, and they didn't want to add hundreds of thousands of new cases—and a hefty dose of embarrassment—to the list. "We can't handle the ones we have already identified," they wailed.[6] They didn't *want* to know who was here that shouldn't be. It seems like they still don't.

That's too bad, since the 9/11 terrorists all came into the United States through the front door. None of them slipped over our border under cover of night. A routine check of expired visas would have allowed us to kick seven of the hijackers out of the country, including ringleader Mohamed Atta. Indeed, traffic police actually stopped Atta and two other hijackers in the months before 9/11. But the cops had no way of knowing either that they were on terror watch lists or that their visas had expired, so the terrorists were released.

Although many of the hijackers came here with visas, a study conducted by *National Review* magazine came to a shocking conclusion: fifteen of the nineteen hijackers should never have been issued visas in the first place. So lax was our review of visa applications that we welcomed fifteen men into our country who clearly should have been kept out—men who went on to smash airplanes into the World Trade Center and the Pentagon.

National Review asked a panel of six experts to review the original visa applications of the 9/11 hijackers. The panel included "four former consular officers, a current consular officer stationed in Latin America, and a senior official at Consular Affairs—the division within the State Department that oversees consulates and visa issuance—who has extensive consular experience."[7]

Every one of the six experts "strongly agreed" that fifteen of the nineteen visas had "managed to slip through the cracks." "Making the visa lapses even more inexplicable," the magazine also reported, "the State Department claims that at least eleven of the fifteen were interviewed by consular officers. Nikolai Wenzel, one of the former consular officers who analyzed the forms, declares that [the State Department's] issuance of the visas 'amounts to criminal negligence.'"[8]

At the risk of nauseating our readers, what follows is a review of the massive gaps in our border protections that permitted these vicious men to enter our country. None of these men swam the Rio Grande. None came from a nation where visas were not required. All came here legally, with temporary visas issued by the INS, deliberately and successfully exploiting the weaknesses in our immigration visa system.[9]

TEN OF THE 9/11 HIJACKERS COULD HAVE BEEN DEPORTED *BEFORE* THEY STRUCK

- Hani Hasan Hanjour, pilot of AA 77: visa expired
- Khalid al-Mihdhar, hijacked AA 77: on a terror watch list; held multiple passports, but nobody picked him up
- Nawaf al-Hamzi, hijacked AA 77: on a terror watch list; visa expired
- Mohamed Atta, pilot of AA 11: arrested and failed to show up in court for driving without a license in Florida
- Satam al-Suqami, hijacked AA 11: visa expired
- Waleed al-Shehri, hijacked AA 11: visa expired
- Wail al-Shehri, hijacked AA 11: here on a doctored passport
- Abdulaziz al-Omari, hijacked AA 11: here on a doctored passport
- Marwan al-Shehhi, pilot of UA 175: visa expired
- Ahmed al-Ghamdi, hijacked UA 175: visa expired

THE HIJACKERS OF AA FLIGHT 77, WASHINGTON TO LOS ANGELES—THE PLANE THAT FLEW INTO THE PENTAGON

Hani Hasan Hanjour, of Saudi Arabia, is believed to have piloted the plane. Hanjour lived in the United States in 1991 and again from 1996 to 1999. In 2000, he was denied a student visa because he lied on his application, failing to reveal his previous stays in our country. But when he applied again, nobody held the earlier lie against him. He was admitted on a student visa but never went to school. His visa had expired at the time of the attack. Hanjour should never have been allowed in the United States. Once he did gain entry, he should have been expelled after failing to show up for school. When his visa expired, he certainly should have been deported. But none of those things happened. Stopped for speeding in Arlington, Virginia, Hanjour simply paid the speeding ticket.

Khalid al-Mihdhar (or Almidhar) of Saudi Arabia, was admitted to the United States on July 4, 2001. He, too, lied about his previous visits to the States, and U.S. officials did not know that he had multiple passports. Suspicious indicators led to his inclusion on a terror watch list in August 2001—but no one moved to pick him up, even though he lived quite openly in New York and carried a Virginia driver's license issued in his real name.

Nawaf al-Hamzi (or Alhamzi), also of Saudi Arabia, was added to the terror watch list in August 2001, but was never picked up. He entered the United States legally in January 2000 after a stay in Malaysia; in July 2000, INS renewed his visa for six months, even though the CIA thought he was dangerous enough to have him followed while he was in Malaysia. By September 11, he had overstayed his visa by nine months, had three different driver's licenses, and was on the terror watch list. But no attempt was made to pick him up.

Salem al-Hamzi (or Alhamzi), of Saudi Arabia, brother of Nawaf, was admitted to the States in June 2001, and was here legally at the time of the attacks.

Majed Moqed, a Saudi Arabian whose identity is in doubt, appears to have entered the United States legally in May 2001.

THE HIJACKERS OF AA FLIGHT 11, BOSTON TO LOS ANGELES—THE PLANE THAT FLEW INTO THE WTC'S NORTH TOWER

Mohamed Atta, of Egypt, piloted the plane. By all accounts the 9/11 ring-leader, Atta was on a terror watch list; he was detained in the United Arab Emirates (UAE) in January 2001 but was released because the United States had not filed any charges against him. But UAE authorities say they warned U.S. officials that Atta was returning to this country, where he had lived off and on for the previous two years. Despite this warning, and despite the fact that he had overstayed his visa in December 2000, Atta was permitted back into the States on January 10, 2001. He left the country for a few weeks in June 2001 and was readmitted on July 19, 2001. The INS even agreed to change his status to that of a trainee and extended his visa. (He was actually sent a change of status notification *a year after 9/11!*) Arrested in Florida for driving without a license, Atta failed to show up for his hearing, which triggered a bench warrant for his arrest; it was never enforced.

It is TRULY an outrage that no one connected the UAE warning with the Florida bench warrant and his frequent INS applications—thereby allowing him to pass freely across our borders time and again.

Satam al-Suqami, a Saudi Arabian, was here illegally at the time of the attacks, after entering the United States in May 2001 for what was supposed to be only a twenty-day period. But no one ever bothered to pick him up—and three months later he became one of the hijackers.

Waleed al-Shehri (or Alshehri), a Saudi Arabian, had also overstayed his visa and should never have been in the United States at the time of the attack.

According to a *National Review* article, Waleed and his brother Wail (see below) "applied together for travel visas on October 24, 2000. Wail claimed his occupation was 'teater,' while his brother wrote 'student.' Both listed the name and address of his respective employer or school as simply 'South City.' Each also declared a U.S. destination of 'Wasantwn.' Further, what should have raised a consular officer's eyebrows is the fact that a student and his nominally employed brother were going to go on a four- to six-month vacation, paid for by Wail's 'teater' salary, which he presumably would be foregoing while in the United States. Even assuming very frugal accommodations, such a trip for two people would run north of $15,000, yet there is no indication that the consular officer even attempted to determine that Wail in fact had the financial means to fund the planned excursion. They appear to have received their visas the same day they applied." [10]

Wail (or Wael) al-Shehri (or Alshehri), a Saudi Arabian, brother of Waleed, obtained a tourist visa in Saudi Arabia. The INS failed to notice, at the time of his admission to this country, that he had doctored his passport by removing pages that indicated he had been to terrorist nations, presumably for training.

Abdulaziz al-Omari (or Alomari), a Saudi Arabian, obtained a tourist visa in his native country in June 2001. But he, too, removed suspicious pages from his passport without attracting notice from the INS.

As the *National Review* reports, al-Omari "claimed to be a student" on his visa application, "though he left blank the space for the name and address of his school. He checked the box claiming he was married, yet he left blank the area where he should have put the name of his spouse. Although he claimed to be a student, he marked on his form that he would self-finance a two-month stay at the 'JKK Whyndham Hotel'—and provided no proof, as required under law, that he could actually do so." [11]

Despite the legal requirement that visa applicants show strong roots in their home country (to give them a reason to return from America), Alomari listed his home address as the "ALQUDOS HTL JED" (a hotel in Jeddah, Saudi Arabia). Alomari didn't even bother filling in the fields asking for his nationality and gender, apparently realizing that he didn't need to list much more than his name to get a visa to the United States. He was right. [12]

HIJACKERS OF UA FLIGHT 175, BOSTON TO LOS ANGELES— THE PLANE THAT FLEW INTO THE WTC'S SOUTH TOWER

Marwan al-Shehhi (or Alshehhi), of the United Arab Emirates, a pilot, overstayed his visa in November 2000, and left the country in December. Despite the fact that he had clearly violated our laws by overstaying his visa, he was allowed to reenter the United States in January 2001, and then again in May of that year.

Fayez Ahmed Rashid Ahmed al-Qadi Banihammad (aka Fayez Ahmed), of the United Arab Emirates, obtained a tourist visa in UAE and entered the States in June 2001.

Ahmed al-Ghamdi (or Alghamdi), of Saudi Arabia, was here illegally at the time of the attack, having overstayed his visa.

Hamza Saleh al-Ghamdi (or Alghamdi), of Saudi Arabia, obtained a visa in his native country.

Mohand al-Shehri (or Alshehri), of Saudi Arabia; his identity is in doubt. After obtaining a tourist visa in Saudi Arabia, he was admitted to the United States in May 2001.

HIJACKERS OF UA FLIGHT 93, NEWARK TO SAN FRANCISCO—THE PLANE THAT CRASHED IN PENNSYLVANIA

Ziad Samir Jarrah, of Lebanon, a pilot. After getting his tourist visa in Germany, he entered and left the United States five times in the two years before the attack, returning for the last time in August 2001.

Saeed al-Ghamdi (or Alghamdi), of Saudi Arabia, lied on his visa application, denying that he had ever previously applied for one, but the Consular Service did not catch the lie. Because he had only a one-way ticket, he was sent to secondary inspection on arrival in the United States, but was eventually admitted anyway.

Ahmed Ibrahim A. al-Haznawi (or Alhaznawi), of Saudi Arabia, was admitted to the United States even though his passport may have had "suspicious indicators," according to the 9/11 Commission Staff Report.

Ahmed Abdullah al-Nami (or Alnami), a Saudi Arabian. Another hijacker whose passport may have had "suspicious indicators," according to the 9/11 Commission Staff Report.

Of these nineteen hijackers, then, five had overstayed their visas and should not have been in the United States on 9/11. Another two had previously overstayed visas, but had been given a pass—despite having flouted the laws of the United States.

No wonder we're outraged, when this is how the State Department and ICE protects our borders!

There were 1.5 million immigrants from the Middle East (from Morocco to Pakistan) living in the United States in 2000, up from 200,000 in 1970. About ten percent—150,000—are here illegally, most because they've overstayed their visas.[13]

And more immigrants from the Middle East are lining up to come to the United States. As the *National Review* wrote, "Even after the terror attacks, in October 2001 the State Department received some 1.5 million entries from the region for the visa lottery, which awards fifty thousand green cards worldwide to those who win a random drawing. Assuming no change in immigration policy, we project that in just the next decade, 1.1 million new immigrants (legal and illegal) from the Middle East will settle in the U.S. Looking forward a little further, within less than twenty years the number of Muslim immigrants and their progeny will grow to perhaps six million."[14]

As suggested by these harrowing stories, it is the ICE inspectors at the port of entry who constitute our first line of defense, by deciding whom to admit and for how long they'll be allowed to stay. Despite the failures of this process to keep out the 9/11 terrorists, in 2001, 573,000 people were *actually* turned away by INS inspectors. If they had turned away nineteen more, they could have prevented 9/11.[15]

The 9/11 stories demonstrate how thoroughly fraud can subvert the screening process. As Jessica Vaughan writes, "By all accounts fraud is rampant in the non-immigrant visa program. Problems include identity fraud, document fraud, counterfeiting, corrupt employees (both American and foreign), and widespread lying and misrepresentation on the part of applicants."[16]

In theory, if an applicant commits fraud on his visa application, he or she can be barred, for life, from entering the United States. But the entire thrust of the regulations governing consular officials discourages the imposition of this penalty. Vaughan writes that "even if the intrepid consular officer has the time and the boss's blessing to pursue and punish fraud, he or she still faces a major obstacle in the form of . . . a full twenty-two pages of daunting legalistic instructions that in practice serve as a powerful disincentive to invoking this section of the law, even in seemingly straightforward cases. Early on, these instructions include a caveat essentially implying that . . . making someone permanently inadmissible because they resorted to fraud in their application is a very mean thing to do." [17]

According to these official guidelines, a consular officer cannot refuse a visa on the grounds of misrepresentation or fraud unless

- he can prove the " 'materiality' or relevance of the misrepresentation to the ultimate decision. The misrepresentation is not considered material if the officer would have denied the application anyway";

- the misrepresentation is something other than "lying about one's identity, address, or previous visa applications," offenses that are "not to be considered material"; (Are they kidding? What else could possibly be considered "material"?)

- his name is not listed in CLASS [the Consular Lookout and Support System]. Unbelievably, if an applicant's name appears here, he cannot be found permanently inadmissible because "it is assumed that the officer would have found out and refused the visa anyway." [18]

Even after a consular official has jumped through all these hoops, he still has to submit a detailed justification for any decision to deny entry—a process so daunting that officials sought to invoke fraud as a preclusion for immigration in only fourteen thousand cases in 2001.

But, even then, the applicant can seek a waiver on the lifetime ban on entry to the United States. And once this waiver application is submitted, the consular official must forward the application to Washington. Vaughan concludes: "It is no wonder that most . . . officers will elect to deal with fraud by simply refusing the applicant who uses fake documents . . . and moving on to the next person. The problem with that approach is that it has no deterrent effect or meaningful consequences for the applicant, and reduces the [nonimmigrant visa] application process to a game of 'Can You Fool the Consular Officer?' "

Fortunately, there is at least one respect in which ICE and the Consular Service have really begun to clean up their acts: the enforcement of student visas.

As noted, hijacker pilot Hani Hanjour came to the United States on a student visa. He never went to school, but no matter: back then, nobody bothered to check.

Since 2003, in a direct response to the utter failure to police student visas that helped to make the 9/11 attacks possible, "more than 10,300 schools and universities . . . with more than 1 million foreign exchange visitors and students" participate in ICE's Student and Exchange Visitor Program, according to ABC News.[19]

The goal of the program is to make sure that those here on student visas are really in school, not just using student visas as an excuse to get into the country. Since the program was initiated, the INS has generated seventy-six hundred leads on potential violations of student visas, and more than eighteen hundred people have been arrested.

The system "now appears to be functioning quite well in helping track those foreigners in the United States on student visas," Brookings Institution senior fellow Michael O'Hanlon said in recent congressional testimony. "Those who overstay visas can be more quickly identified and located."[20]

But efforts to stop other visitors from overstaying their visas are not going very well at all. Homeland Security, under the US-VISIT program launched in 2004, requires that people who are applying to enter the States be fingerprinted and photographed in their home countries before they arrive here. As they enter, their fingerprints are scanned to be sure that they are indeed the same person who initially applied for the visa.

One hundred and fifteen airports and fourteen seaports have the US-VISIT program in place and meticulously fingerprint those who enter the country through their portal. But only twelve airports and two seaports bother fingerprinting those who leave. As *Newsday* noted, "Without exit checks the system can't flag people who linger after their visas expire, and there's currently no timetable for expanding those checks beyond the existing pilot program."[21]

On August 1, 2006, the Department of Homeland Security proposed a rule to fingerprint "millions of legal permanent residents with green cards when they reenter the country after traveling abroad." But that's not the problem. The problem is visa overstays, and neither the current system nor the proposed one will do any good in dealing with this key issue.[22]

While fingerprinting is a useful tool to help turn away those who might turn out to be terrorists, it could also be used to stop people from overstaying their visas. Tracking the names of aliens who entered the country, and whether they left on time, would give us a list of illegal immigrants still living in the United States—their names, photographs, fingerprints, along with their stated intention to live in a certain place and do a certain kind of work. Suddenly, finding these offenders looks less like an insurmountable obstacle—and more like an urgent imperative.

But if we continue to check arrivals at 115 airports and departures only at twelve, we will never be able to subtract the list of those who left from those who arrived to find out who is overstaying his visa. The Department of Homeland Security and ICE have no plans to expand the checks on departures, so it cannot be used as an effective tool to regulate our borders, which will remain as open as a sieve so long as the Department maintains this approach.

But the sieve is even more porous than it would appear at first glance. Of the 33 million foreigners who enter our borders legally each year, about half come without any visa at all under the Visa Waiver Program, which lets the citizens of twenty-eight countries enter without visas if they are staying for ninety days or less. It makes no difference if they are coming for tourism or business.[23]

Many of these visa waiver countries, of course, are our traditional allies and friends in Europe. As members of the European Union, however, these nations have scant ability to police their own borders. The free flow of population and workers that is the cornerstone of the EU makes it impossible for a visa waiver country like Britain to stop a flood of about 250,000 immigrants a year from the European continent. As it stands today, immigrants from terrorist-friendly countries can shop for the easiest point of entry into Europe, migrate easily from there to a visa waiver country, and then come to the United States without any visa at all. No consular scrutiny, no fingerprints, no photographs—just a passport.

If the European Union proceeds with its plans to admit Turkey, with its large Islamic population and its porous borders, the visa waiver program will present us with an even more significant security problem.

THE TOTAL NUMBER OF ENTRANTS INTO THE UNITED STATES IS QUICKLY OUTSTRIPPING THE NUMBER WHO BRING VISAS WITH THEM[24]

Table 1. Non-Immigrant Visas (NIVs)

Year	Number of Visas Issued	Number of Admissions *
1985	5,796,034	9,540,000
1990	6,034,253	17,574,000
1995	6,181,822	22,641,000
2000	7,141,636	33,690,000
2001	7,588,775	32,824,000

* Numbers are approximate.
Source: U.S. State department and INS.

Easy access for terrorists is not the only problem with illegal immigration; ordinary criminals are just as likely to exploit the system.[25]

CRIME: THE COST OF ILLEGAL IMMIGRATION

- 27 percent of federal prison inmates are illegal immigrants

- We spend more than $1 billion per year to keep illegal immigrants in federal prisons

- About 300,000 illegal immigrants are in state prisons

- In Los Angeles, 95 percent of outstanding homicide warrants are for illegal immigrants

Though the problem of illegal immigrants in prison is huge, even worse is the number of immigrant criminals who are at large. As the Center for Immigration Studies (CIS) points out, in Los Angeles two-thirds of all fugitive felony warrants are for illegal aliens.

A 1995 California Department of Justice study revealed that illegal immigrants form the core of some of the state's most despised criminal gangs, including 60 percent of the notorious 18th Street Gang, believed to number twenty thousand or more. The report describes the gang and its ally, the Mexican Mafia, as "the dominant force in California prisons" and says that their specialties are "complicated drug distribution schemes, extortion, and drive-by assassinations."[26]

Why can't we track down illegal immigrants, lock them up, check their status, and then deport them? The answer: a combination of limited resources, civil liberties, and political correctness.

Let's start with catching illegal immigrants. ICE has two thousand officers charged with finding and deporting illegal immigrants. How many illegal immigrants are there on the streets of America? Eleven million.[27] With the ICE so vastly outnumbered, one obvious solution would be to call on local police for support. In most cities, however, police officers aren't allowed to use immigration violations as a grounds for arrest.

As the Center for Immigration Studies reports, in many cities "the police cannot use . . . immigration status [to apprehend criminals]." If a Los Angeles Police Department officer arrests an illegal immigrant, "it is the officer who will be treated as a criminal by his own department—for violating the LAPD's rule against enforcing immigration law. The LAPD's ban on immigration enforcement is replicated in immigrant-heavy localities across the country." New York, Chicago, Austin, San Diego, and Houston are among dozens of cities that have so-called sanctuary policies, regulations "which generally prohibit a city's employees, including the police, from reporting immigration violations to federal authorities."[28]

WHERE THE COPS DARE NOT GO: SANCTUARY CITIES THAT BAR COPS FROM ENFORCING IMMIGRATION LAWS

Phoenix, Arizona	San Diego, California
Los Angeles, California	San Francisco, California
Long Beach, California	Denver, Colorado
Sonoma County, California	Miami, Florida

Chicago, Illinois
Cambridge, Massachusetts
Baltimore, Maryland
St. Paul, Minnesota
Jersey City, New Jersey
Trenton, New Jersey
Brentwood, New York
Central Islip, New York
Farmingville, New York
New York, New York
Peekskill, New York
Riverhead, New York

Charlotte, South Carolina
Raleigh, North Carolina
Winston-Salem, North Carolina
Portland, Oregon
Austin, Texas
Dallas, Texas
Houston, Texas
Fort Worth, Texas
Salt Lake City, Utah
Fairfax County, Virginia
Madison, Wisconsin

Houston is one of the worst cases. According to Craig Ferrell, general counsel for the Houston Police Department, officers in the city "can't legally question individuals solely on the suspicion they are in the country illegally." [29] Why not? Ferrell's alibi is that "police need immigrants to feel comfortable talking to authorities and not fear deportation. . . . There are lots of people with a questionable immigration status," he says. "They are witnesses to crimes. It's important to have assistance from law-abiding people to solve crimes and keep the city safe." [30]

But Ferrell and his ilk don't ask the other question: how many of those who *commit* crimes are illegal immigrants? An immigration violation can be a useful—and entirely legitimate—way to catch a criminal and hold him while the police amass evidence of other crimes the detainee may have committed.

So why can't cops snare illegal immigrants? Officially, the answer is that immigration regulations are a federal matter, so local police have no business enforcing them. But there's another, real-world reason: police departments are reluctant to burn their bridges to the minority communities where the illegals live, and local politicians don't want to lose votes just to enforce a federal law.

Yet, if we were to let police enforce immigration laws, the results could be startling. In the mid-1990s, Congress authorized local police forces to help enforce immigration laws, but it was only after 9/11 that any locality sought training under the program. Orange County, California, was one of the first. As *National Review* reports, the police there "actively seek out [illegal immigrant] criminals rather than waiting for a law-breaker to come to them. In these gang-saturated neighborhoods . . . the only tool that a law-enforcement officer may have for getting an illegal gang-banger off the street is his immigration status; trying to build a case for armed robbery, say, may be futile. Moreover, if an investigator has only enough evidence to detain someone briefly for questioning about a crime, but not yet probable cause to arrest him, a quick check of the immigration database may provide grounds to arrest him rather than let him get away."[31]

The problem of catching visa overstays is particularly difficult because of the antiquated data processing of the new agency, the United States Customs and Immigration Service (USCIS), which is the other half of the old INS. According to a report by the Department of Homeland Security's inspector general, "USCIS faces the continuing challenges of . . . modernizing its technology." Its systems are "primarily manual, paper-based, and duplicative, resulting in an ineffective use of human and financial resources to ship, store, and track immigration files. Adjudicators use multiple and non-integrated IT systems to perform their jobs, which reduces productivity and data integrity."[32]

Worse, the files for immigration cases are shipped multiple times across the country in the course of any prosecution, careening from district offices to a Chicago lockbox, to the National Benefits Center, back to the district office, off to the National Records Center in Missouri, then to the relevant service center, before making it back to the district office and finally back to the Missouri National Records Center. In fiscal 2004, the USCIS's budget for record management alone was more than $80 million! One service center racked up costs at $400,000 just for copiers and copy paper and $2 million for mailing costs. Also, the "USCIS pays a contractor about $90 million a year for filing and data entry. Data entry is a particularly inefficient process, as the same information is rekeyed multiple times when files are transferred among USCIS locations."

Welcome to the Stone Age!

ARE WE CATCHING UP? NO WAY!

- Number of illegal immigrants in U.S. today: 11.5 million
- Current backlog of immigration cases: 1.5 million
- Estimated number of illegal immigrants in state prisons: 300,000
- Number of agents charged with finding and deporting illegal immigrants already here: 2,000
- Number of beds in detention facilities: 27,500

In 2002, President Bush initiated a $500 million program to eliminate the backlog of immigration and naturalization applications. By the end of 2004, the backlog totaled about 1.5 million cases. Some progress!

Even if, by some miracle, we should catch a given violator who's over-stayed his visa, chances are we'd have no place to hold him. Our detention cells are so inadequate, our local jails so overcrowded, that immigration authorities practice a kind of triage, pursuing only the worst cases. Top priority goes to catching aliens "who have been deported after having committed a crime and then illegally reentered the country." Next come aliens who have reentered the country after being deported. The third priority is illegal aliens who have committed a crime. Nobody else, the Center reports, has much to fear from immigration authorities. "An illegal alien who has merely been arrested 14 times for robbery, say, without a conviction will draw only a yawn from an ICE district director." [33]

In his 2007 budget, President Bush asked Congress for 4,000 more beds in detention facilities, pushing the total up to 27,500. But even this increase is laughably inadequate. With about 1.5 million people in prison, and an estimated one-third of them illegal immigrants, how does a gesture of 4,000 extra beds address a pool of 500,000 illegal immigrant inmates who will be coming out of prison when their sentences are up?

Deporting an illegal immigrant is no easy thing to do. The American Immigration Lawyers Association (AILA) has helped win "an elaborate set of trial rights for criminal aliens that savvy attorneys can use to keep them in the country indefinitely." [34] As one probation officer told the Center for

Immigration Studies, "A regular immigration attorney can keep you in the country for three years, a high-priced one for ten."[35]

But the disaster doesn't end here. Even if we catch an illegal immigrant and succeed in deporting him, the Center for Immigration Studies reports that the "lack of resources . . . derails the conclusion of the deportation process."[36] Though ICE may win the right to deport an illegal alien, it "rarely has the manpower to do so." The Center reports that, in the early 1990s, only fifteen INS officers were responsible for the deportation of 85,000 aliens in New York.

So ICE issues what are known as "run letters," which advise the alien that he has to report in a month or so to be deported. ICE sends the letters in the full expectation that the recipient will take the hint and leave on his or her own. Eighty-seven percent do. But ICE's limited manpower and money is overtaxed by dealing with the ones who don't.

The worst horror is that the massive number of illegal aliens in our prisons has largely stopped efforts to deport illegal immigrants who have committed crimes, been caught, tried, and sentenced, served their time in prison, and are about to be released. Even here, ICE can't assure us that these ex-cons will leave the United States.

In 1988, the INS started the Institutional Hearing Program so that it could complete deportation hearings while an alien detainee was still in jail. The idea was that the prisoner would be kicked out as soon as his sentence was up. But the Center for Immigration Studies reports that "the process immediately bogged down due to the magnitude of the problem—in 2000, for example, nearly 30 percent of federal prisoners were foreign born. The agency couldn't find enough pro bono attorneys to represent criminal aliens (who have extensive due process rights in contesting deportation), and so would have to request continuance after continuance for the deportation hearings. Securing immigration judges was a difficulty as well."[37]

The result was incredible: in 1997, the INS didn't know where a third of the foreign-born inmates went after they were released from prison. They were right back on the streets, in the United States.

As Americans for Legal Immigration PAC president William Gheen notes, "Illegal aliens have been walking out of American prisons after serving their time at taxpayer expense without being deported. Our government can't or won't find the hundreds of thousands of known felon illegal aliens walking America's streets tonight, much less stop the new felons coming in tonight across our unsecured borders."[38]

According to Chelsea Schilling, who has written for WorldNet Daily about gang crime by illegal aliens, "Last year, officials of the House Judiciary Committee said that U.S. immigration officers and police are not always on the same page. Police do not always inform immigration authorities about arrests of undocumented aliens, and immigration officers are often too late to identify the aliens before they are released on bail."

For example, New York's arm of the Department of Homeland Security is only interviewing 40 percent of foreign-born inmates at Rikers Island, "a failure that puts criminal aliens back on the streets instead of deporting them," according to the *New York Post*.[39]

Despite all the obstacles, ICE has been "expelling" a larger number of people each year than ever before. As the following chart indicates, the record of deportations has improved considerably, particularly in the Clinton years:

DEPORTATIONS BY YEAR[40]

1991	33,189	1998	173,146
1992	43,671	1999	181,072
1993	42,542	2000	186,222
1994	45,674	2001	178,026
1995	50,924	2002	150,542
1996	69,680	2003	189,368
1997	114,432	2004	202,842

It is estimated that about half of those who *were* deported had committed crimes other than their immigration violations.[41] (In addition to those who were deported, the government says that slightly more than a million left "voluntarily.")

Particularly encouraging are the Fugitive Operations teams set up by ICE, now operating in forty-five areas throughout the country. These units are specifically designed "to find, arrest, and place into removal proceedings aliens who have been ordered by an immigration judge to leave the country, but have failed to comply—thus making them fugitive aliens."[42]

The ICE Fugitive Operations teams have arrested 52,000 illegal aliens since their inception in 2003, including 22,669 who "had convictions for crimes that include homicide, sexual assault against children, robbery, violent assault, narcotics trafficking, and other aggravated felonies and crimes of moral turpitude."[43] With eleven million illegal immigrants in the United States, all these numbers seem just a drop in the bucket.

However, when an election is on the line, the federal government can cut all the red tape and handle immigration processing in record time. Facing a tough battle for reelection in 1996, President Clinton set up the Citizenship USA program to expedite the processing of citizenship applications from immigrants—just in time for them to vote in the elections.

Spurred by welfare reforms, which limited certain benefits to citizens, massive numbers of immigrants sought citizenship under the Expedited Review Program and hundreds of thousands became citizens. But an audit showed that 99 percent of applications in New York and 90 percent in Los Angeles had errors. "As a result, tens of thousands of aliens with criminal records, including for murder and armed robbery, were naturalized."[44]

If only the urgency of pursuing illegal aliens who overstayed their visas had had a comparable urgency in either the Clinton or the pre-9/11 Bush administrations, the seven hijackers who were here illegally might have been apprehended.

ACTION AGENDA

The first question is whether we should try to stop all immigration from countries that sponsor or promote terrorism. *National Review,* America's leading conservative publication, says this idea is not really worth debating. "Even after September 11," it notes, "not a single member of Congress proposed cutting off Middle Eastern immigration. Congress would never single out one region of the world for exclusion from green cards." The magazine writes that "it is politically inconceivable, in our equality-obsessed society, that we would ever return to the days prior to 1965 in which some regions of the world were allotted fewer green cards than others."[45]

Political correctness be damned. We say: in our view, the United States should bar immigration from all countries that sponsor terrorism or harbor terrorists. Here's the list:

TERROR SPONSORS

Cuba	North Korea	Syria[46]
Iran	Sudan	

SAFE HAVENS

Mali	Cyprus	Syria
Mauritania	Chechnya	Iran
Somalia	Afghan-Pakistan border	Colombia (certain regions)
Philippines (southern region)	Lebanon	Venezuela[47]
Indonesia	Yemen	

Why should the United States subject its residents and citizens to the risk of terrorism by permitting people from these nations into our country? After all, in many respects, we treat terror-sponsoring or terror-harboring nations differently from the rest. Our policy toward them is different; we handle our relations with them differently.

It is by no means unusual to have one set of rules for one country and another set for another one. The Visa Waiver Program gives an explicit preference to visitors from twenty-seven countries, allowing their citizens to come here without applying for visas in advance. Recognizing that these nations pose little threat, the United States doesn't want to inconvenience their citizens who wish to come here. Why not treat immigrants from nations that sponsor or harbor terrorists differently from visitors from other lands?

Some say we should allow immigration from these nations, but should scrutinize the immigrants more carefully. This is a fantasy. How can the United States consular officials hope to know if a person is a potential terrorist from a nation with which we do not have diplomatic relations, like Iran? What database are we supposed to use? Who would do the checking? Are we to rely on foreign government agencies for advice? Should we take at face value the documentation we get from the applicants?

It may be harsh to bar citizens of certain nations to come to the United States—especially when they may be family members of U.S. citizens or permanent residents. But the threat of terrorism, too, is harsh. The United States is under no moral, legal, constitutional, or ethical imperative to open our borders to the people of a particular nation. Beyond any future limitations, we must immediately place a dramatically higher priority on catching those who have overstayed their visa departure dates and who are already here illegally. This is no small matter. Patrolling the Mexican border as intensively as possible would not reduce the potential terrorist population in the United States by a fraction of what good visa enforcement would achieve. As we've noted, most terrorists don't enter this country from Mexico. They enter through JFK airport and the other legal ports of entry into the United States.

Right now, roughly four or five million people who overstayed their visas are living illegally in the United States. Their willingness to flout our laws in this regard clearly suggests that they may be lawless in other respects—and painful recent experience reminds us that those respects may include terrorism.

But our resources are limited. So let's at least target those visa violators who come from terror-sponsoring or terror-harboring countries! This is not racism—it's due diligence. If the United States Department of State feels comfortable declaring, after extensive study, that certain countries sponsor or harbor terrorists, we should feel comfortable detaining people from these nations who have violated their agreement to leave our shores after a certain period of time.

Right now, ICE gives top priority to criminals who have left prison and are here illegally. This focus is obviously important. But let us not forget the lesson of 9/11: a terrorist can kill a lot of people at once. It is vital that we raise to the same level of importance the pursuit of visa violators who come from terror-sponsoring nations.

THE UNITED NATIONS

The United Nations, created to be mankind's best hope for peace, has become the front for the largest corruption scandal in history: its Oil-for-Food Program. This scandal—which involved the theft of almost $14 billion and corruption by high-level diplomats, UN employees, and advisers, as well as over two thousand companies from more than sixty different countries—has left the United Nations tarred and tainted, perhaps beyond repair. More than any other episode, it illustrates the corruption that pervades the international community, many of whose members are less rigorous in their enforcement of ethical standards than the United States, condoning corruption and making a national practice of looking the other way.

But Oil-for-Food is only one of the scandals that have sapped the reputation of the United Nations. Drug dealing among its bureaucrats, incredible overruns in its budget for renovation of its headquarters, sexual abuse by its peacekeeping forces, fraud in its procurement department, and the mockery its Human Rights Commission has made of the enforcement of the rights of man—all of these tell the story of a dysfunctional and corrupt organization.

The United Nations is like a government with a bicameral legislature—the General Assembly and the Security Council—and an executive branch in the secretary-general. But there's no judiciary branch. The World Court adjudicates cases brought to it, and chases criminals against humanity, but does nothing to police the operations of its host body, the United Nations.

Because of this lack of judicial oversight, there is no method of deterring

SFS0023099
PROVIDENCE ACADEMY HS LIBRARY
106 TROY STREET
SENECA FALLS, NY 13148

or punishing malfeasance within the UN. Even in a horrendous scandal like Oil-for-Food, where the program director himself took $150,000 in bribes, the UN can mete out no punishment beyond dismissal. There is no watchdog, no overseer, no prosecutor, and no court with jurisdiction over such activity.

The UN still functions as if the General Assembly and the Security Council were the entire organization. But the UN has spawned a gigantic bureaucracy to implement the collective will of these bodies. It administers ninety thousand peacekeeping troops in fifteen different nations, and it has to provide them with full logistic support. As the world's only global administrative body, it is frequently saddled with missions like the Oil-for-Food program. The scandals that are ripping apart the United Nations today only demonstrate how insufficient its current structure is to deal with these challenges.

While serving as interim U.S. ambassador to the UN, John Bolton outlined a vigorous program of management reform of the world body, making dozens of specific suggestions for budget caps, management audits, strengthening of oversight, and other controls. He formed a coalition of fifty nations that contribute 87 percent of the UN's budget (led by the United States, the largest donor, supplying 22 percent). But his reforms met opposition from a group of seventy-seven nations who contribute only 12 percent of the UN's revenues—a group that dismissed Bolton's reforms as unnecessary measures that would hamper the organization's effectiveness. With no Democratic support, Bolton resigned his post after the November 2006 election.

The Oil-for-Food debacle, in its gory details, shows how a combination of appeasement, self-dealing, lying, nepotism, and cronyism underscore the value systems of the United Nations and its European advocates and members. More than an isolated scandal, it offers a glimpse of how fundamentally dishonest some of our purported allies in Europe really are and how they use their ideological posturing as a way to cover their financial chicanery.

OIL FOR FOOD

The sheer magnitude of the larceny in the Oil-for-Food Program and its political implications are staggering. Worse, they suggest that French, German, Russian, and Chinese opposition to a tough policy in Iraq, including their criticism of the U.S.-UK invasion, was motivated not by principle or a desire for peace, but by greed, extortion, and bribery. Worse still, it casts doubt on the integrity of the highest levels of the executive branch of the United Nations—going up the ladder to the recently departed secretary-general himself, Kofi Annan.

The principle of the Oil-for-Food Program was that Iraq would be permitted to sell its oil on the world market and then use the money to buy "humanitarian" supplies. The United Nations was supposed to oversee the process to make sure that Iraq did not buy any products that might help it arm itself and to see to it that all the humanitarian products reached the needy people of Iraq. It failed on both counts.

The origins of the Oil-for-Food Program can be found in the wake of Saddam Hussein's invasion of Kuwait in August 1990. The United Nations reacted by banning its members from trading with his regime until it had disarmed to the satisfaction of the United Nations. Saddam refused to submit, and the sanctions lasted well into the 1990s.

Saddam soon began his campaign to overturn the sanctions, working both overtly—citing the plight of the children of Iraq who were unable to get needed food and medicine—and covertly—by bribing people in positions of power at the United Nations and Western European governments.

The sanctions did not bar the importation of food or medicine, but Saddam claimed he couldn't get the necessary foreign reserve to buy them unless he was allowed to sell oil. So the global community, led by then-UN secretary-general Javier Perez de Cuellar, demanded that Saddam be allowed to sell strictly limited quantities of oil to fund his purchases of food and medicine; the result was the Oil-for-Food Program.

At first Saddam made a show of resisting the program, forcing the West to beg him to allow the sale of his oil in return for relief supplies for his people. He eventually submitted, allowing the West to claim victory when in fact it was Saddam himself who had scored the key triumph.

The Oil-for-Food Program was created by UN Resolution 986, passed on April 14, 1995. But Saddam's initial negative posture toward the program gave him negotiating leverage over the details of the program, allowing him to influence its final, flawed design. The Clinton administration was so anxious to get the program started—and thus relieve international pressure to lift the sanctions entirely—that it failed to insist on safeguards that would have prevented the wholesale larceny that was to follow.

As Claudia Rosset has written in *Commentary,* Saddam "was given the right to negotiate his own contracts to sell Iraqi oil and to choose his own foreign customers. He was also allowed to draw up the shopping lists of humanitarian supplies—the 'distribution plans'—and to strike his own deals for these goods, picking his foreign suppliers. The UN also granted Saddam a say in the choice of the bank that would mainly handle the funds and issue the letters of credit to pay these suppliers; the designated institution was a French bank now known as BNP Paribas."[1]

But one detail of the deal, more than any other, opened the door to corruption. The costs of the program—of the aid itself and the UN supervision—were to be paid by a 2.2 percent commission on each barrel of oil sold, with an additional 0.8 percent collected to finance the UN weapons inspection program in Iraq. Almost from the start, then, the UN had given itself an incentive to permit Iraq to sell as much oil as possible. Over the life of the Oil-for-Food Program, the UN collected $1.9 billion in commissions on the sale of Iraqi oil.[2]

The program was to be overseen by the Security Council, in practice by its five permanent members. As subsequent investigations would reveal, however, three of the five—France, Russia, and China—were deeply involved in promoting Iraq's oil industry, and its officials were heavily implicated in the kickbacks that flowed from the program. Indeed, Saddam assured lax oversight by the Security Council by awarding $19 billion in oil contracts to Russia and another $4.4 billion to France.[3]

Only the United States and the United Kingdom appeared to show much concern over preventing Oil-for-Food revenue from being used to import "dual use" equipment, which could help in the manufacture of weapons.

In October 1997, the new secretary-general of the United Nations, Kofi Annan, appointed a longtime UN bureaucrat, Benon Sevan, to run the Oil-for-Food Program. Sevan proved a very, very poor choice.

Despite the Security Council's theoretical supervision of the program, the secretary-general himself actually handled its administration through Sevan. As Rosset reports, "the Secretariat was the keeper of the contract records and the books, and controller of the bank accounts with sole power to authorize the release of Saddam's earnings to pay for imports to Iraq. The Secretariat arranged for audits of the program, [and] was the chief interlocutor with Saddam."[4]

Although the Oil-for-Food Program was fairly open at the start, by 1998, its second year, it was wrapped in secrecy. Most of the key information concerning the program was labeled as "proprietary" and kept from the public. "There was no disclosure of such basic information as the names of individual contractors or the price, quality or quantity of goods involved in any given deal," Rosset writes. "Instead the [UN] Office of the Iraq Program released long lists representing billions of dollars in business but noting only the date, country of origin, whether or not the contract had been approved for release of funding, and highly generic descriptions of goods."[5]

Rosset notes that the secrecy coincided with the year that Saddam's government "may have begun covertly sending gifts of oil to Sevan himself by way of a Panamanian firm. . . . It was also the year in which the UN terminated a contract with a UK-based firm, Lloyd's Register, for the crucial job of inspecting all Oil-for-Food shipments into Iraq and replaced it with a Swiss-based firm, Cotecna Inspections, with ties to Kofi Annan's son Kojo."[6]

OIL-FOR-FOOD = BRIBERY

- France: $4.4 billion in oil contracts
- Russia: $19 billion in oil deals
- UN: $1.9 billion in oil commissions
- Kojo Annan (son of Kofi): $10 million contract to Cotecna (his employer)
- Alexander Yakovlev of Russia, senior UN procurement officer: $1 million
- Benon Sevan (administrator of the Oil-for-Food Program): $150,000 cash

Forty countries and 2,250 companies paid bribes to Saddam under Oil-for-Food. George Galloway, a member of the British Parliament, was accused of getting rights to 18 to 20 million barrels of oil—which, if true, could easily have run into the millions of dollars. India's foreign minister, Natwar Singh, and the governing Congress Party steered the rights to four million barrels of oil to a Swiss energy company. Why? We don't know. France's former UN ambassador Jean-Bernard Merimee got $165,725 in oil allocations from Iraq. Even the Vatican was in on it: Rev. Jean-Marie Benjamin, assistant to the Vatican secretary of state—who fought to have the sanctions on Iraq lifted—got oil allocations under Oil for Food.

Indeed, Saddam Hussein was able to run rings around the United Nations management and, ultimately, to pay off enough key officials to make them complicit in his fraud. Once he was allowed to sell oil again, he realized that illegal oil looks about the same as legal oil, and proceeded to rake in a fortune, smuggling oil out through Turkey, Jordan, and Syria almost untraceably. When Iraqi oil started showing up on global markets, who could tell if it was lawful or not? Saddam made almost $12 billion this way, selling oil outside the parameters of the Oil-for-Food Program.

He also made $1.8 billion by forcing companies that sold products and allowable services to Iraq to pay a bribe or kickback for the privilege—another $1.5 billion in kickbacks on humanitarian products.

To grease his path, Saddam corrupted the entire world organization, bribing the director of the Oil-for-Food Program and political leaders in Britain, France, Germany, China, and, especially, Russia to support the program and encourage the lifting of all sanctions. These bribes, which went to 270 separate individuals, were paid in the form of oil vouchers, permitting the recipients to buy Iraqi oil at a discount and then sell it at the higher market price, pocketing the difference.

Saddam may even have attempted to influence Kofi Annan himself. The United Nations hired a firm with which the secretary-general's son, Kojo Annan, was affiliated to inspect cargo shipments to Iraq to prevent weapons-making materials from entering the country. The Swiss firm, Cotecna Inspections, was paid $10 million a year for its services.[7]

As Claudia Rosset has written in *National Review,* however, "as Oil-for-Food worked in practice, there were two glaring flaws." The first was the UN's willingness to allow Saddam to choose buyers for his oil himself, and to give him equal discretion over the suppliers from whom he bought goods. This gave Saddam free rein to negotiate kickbacks, hide imports of prohibited weapons and materials, and extend bribes to the international community. "The other problem was the UN's policy of treating Saddam's deals as highly confidential, putting deference to Saddam's privacy above the public's right to know."[8]

As corruption in the Oil-for-Food Program became more widespread, whistleblowers began to appear. Many were ignored, sidetracked, or fired. One was Paul Conlon, who lost his job in 1995 as a staffer on the UN Iraq sanctions committee. As Eric Shawn reported for Fox News, "Conlon said he repeatedly warned his bosses that Saddam had undermined the sanctions effort with the help of questionable companies that sprang up to provide 'humanitarian' assistance.'" Conlon says that "most of these trading partners that began to turn up on our doorstep with their applications for export permits were very obscure, unknown little companies in very obscure locations."[9] He points to water-purification chemical permits extended to Iraq: "Any kind of chemical whatsoever could have been in those drums."[10]

In the meantime, pressure from global public opinion mounted for an increase in the scope of the Oil-for-Food Program. While the United States and Britain resisted the impetus, Russia, China, France, and Annan himself pressed for expansion of the program. All, it should be noted, had an overt or covert stake in this expansion. Even after Saddam threw weapons inspectors out of Iraq, the program was expanded from a cap of $4 billion annually on oil sales to more than $10 billion per year. Eventually, by 2000, all limits on sales were suspended. Now Saddam could sell whatever he wanted—and, with no inspectors on the ground in Iraq, he was free to use the funds however he liked.

As Rosset observes, the Oil-for-Food Program was a fruitful cover for corruption. "Saddam would sell at below-market prices to his handpicked customers—the Russians and the French were special favorites—and they could then sell the oil to third parties at a fat profit. Part of this profit they would keep, part they would kick back to Saddam as a 'surcharge,' paid into bank accounts outside the UN program in violation of UN sanctions."[11]

The corruption in the program reached staggering proportions, according to an inquiry eventually commissioned by Kofi Annan and headed by former Federal Reserve Board chairman Paul Volcker.

The corruption began at the top. Ryan Balis of the National Center for Public Policy Research described how Volcker accused "Benon Sevan, the former director of Oil-for-Food in Iraq . . . of accepting nearly $150,000 in cash bribes from December 1998 to January 2002 to allocate oil sales from Iraq to a favored company with ties to former UN secretary-general Boutros Boutros-Ghali."[12] Annan's son, Kojo, was accused of lying about his work with the Swiss contractor hired to monitor shipments to Iraq under the program. Other Volcker accusations include the following:

- that Saddam collected $1.8 billion in illicit kickbacks from companies,

- that 40 countries and 139 companies paid bribes to Saddam to buy his oil,

- that Hans Von Sponeck, the former humanitarian coordinator in Iraq, "solicited money from companies seeking to do business" there,[13]

- that Jean-Bernard Merimee, France's former UN ambassador, received $165,725 in commissions from oil allocations awarded him by the Iraqi regime (Merimee is now under investigation in France),

- that Roberto Formigoni, president of the Lombardi region in Italy, received oil allocations under the program, and

- that the Rev. Jean-Marie Benjamin, a priest who once worked as an assistant to the Vatican secretary of state and became an activist pushing for the lifting of Iraqi sanctions, got oil allocations as well.[14]

Volcker stressed that the United Nations had plenty of notice about the corruption of the program but did nothing to stop it. In 2002, Saddam publicly said he would charge a fifty cent per barrel surcharge on Iraqi oil over and above the sales price in the Oil-for-Food Program—a totally illegal step the UN completely ignored.

As Volcker pointed out, "Corruption of the program by Saddam and many participants could not have been nearly so pervasive with more disciplined management by the United Nations." [15] His report also faulted the Banque Nationale de Paris, which issued letters of credit for the program but never shared with the UN the "firsthand knowledge it acquired of the true nature of financial relationships that fostered the payment of illicit surcharges." [16] As Volcker noted, "There were provisions in the program and in its management and oversight that should have permitted it to be caught." [17]

It wasn't enough that Saddam corrupted the sale of Iraqi oil, paying bribes in oil barrels to top UN and European officials or demanding kickbacks and surcharges from oil companies in return for the business. He also insisted on kickbacks from companies that supplied the goods for humanitarian relief in Iraq. Volcker found that sixty-six UN member nations gave Saddam illegal kickbacks—often disguised as "transportation fees" or "after-sales services charges"—and the total of worldwide companies implicated in bribery eventually topped two thousand. [18]

The corruption was ubiquitous. When the United Nations and the U.S. Defense Contract Audit Agency (DCAA) reviewed 759 of the largest contracts under Oil-for-Food, it found that Saddam had overpaid for goods and services in almost half the cases. The overcharges, which averaged 21 percent, totaled $656 million—an amount likely divided between the company and Saddam's regime. In its report, the DCAA noted that "some items of questionable utility for the Iraqi people (e.g., Mercedes Benz touring sedans) were identified." [19]

When the United States and the United Kingdom began to investigate the program and call for closer examination of a number of the relief contracts, "both Sevan and Annan complained publicly and often about these delays, describing them as injurious to the people of Iraq and urging the Security Council to push the contracts through faster." [20]

The *Wall Street Journal* was the first to blow the whistle on the corruption in the Oil-for-Food Program, reporting on May 2, 2002, that "against the resistance of Russia, France, China, and the UN Secretariat, the U.S. and Britain had been trying to put a halt to the kickbacks through an elaborate system to enforce fairer pricing—but with only limited success. Sevan, clearly aware of the scam, said he had 'no mandate' to stop it." [21]

Once faced with increasing evidence of corruption, however, the Security Council actually *loosened* its scrutiny of the program, voting on May 14, 2002 to give direct power to the secretary-general to approve all humanitarian contracts. Annan quickly expanded the definition of humanitarian to include construction, industry, labor and social affairs, the Board of Youth and Sports, information, culture, religious affairs, justice, finance, and the Central Bank of Iraq, as well as other areas having nothing to do with food or medicine. At the end of 2002, with war looming in Iraq, Annan approved a $20 million grant of Oil-for-Food funds to pay for an "Olympic sport city" and $50 million for Saddam's Ministry of Information, his propaganda factory.[22]

It wasn't until after the United States led the occupation of Iraq that the full extent of the corruption of the Oil-for-Food Program became apparent, as auditors from the newly formed Iraqi Governing Council began examining the records of Saddam's government. The revelations were horrendous. News reports identified 270 people in fifty countries who were said to have received vouchers to purchase oil from Saddam Hussein, oil they could resell at a big profit on the global market.

The list included former French interior minister and Chirac intimate Charles Pasqua, British MP George Galloway, Indonesian president Megawati Sukarnoputri, Russia's quasi-fascist political figure Vladimir Zhirinovsky (who had run against Yeltsin and lost), the Russian government itself, the Communist Parties in Russia and other former satellite countries, the Russian Orthodox Church, and several Russian oil companies.[23] In short, Oil-for-Food might have been better named Free-for-All.

Even al Qaeda got in line for Saddam's goodies. *Commentary* reports that "two firms doing business with Saddam through Oil-for-Food were linked to financier Ahmed Idris Nasreddin, now on the UN's own watch list of individuals 'belonging to or associated with al Qaeda.'"[24]

Marc Perelman of the *Jewish Daily Forward* detailed the links between the Oil-for-Food Program and al Qaeda. "One link ran from a UN-approved buyer of Saddam's oil, Galp International Trading Corporation, to a shell company called ASAT Trust in Liechtenstein, linked to a bank in the Bahamas, Bank Al Taqwa. Both ASAT Trust and Bank Al Taqwa were designated on the UN's own terror-watch list . . . as entities 'belonging to or affiliated with al Qaeda.'"[25]

Perelman goes on to point out that "this Liechtenstein trust and Bahamian bank were linked to two closely connected terrorist financiers, Youssef Nada and Idris Ahmed Nasreddin—both of whom were described in 2002 by [the U.S.] Treasury as "part of an extensive financial network providing support to al Qaeda and other terrorist-related organizations."[26]

The second link Perelman uncovered was between a Swiss-based company, Delta Services, which Saddam had picked as one of his oil purchasers. Delta "was a subsidiary of a Saudi Arabian firm, Delta Oil, which had close ties to the Taliban."[27]

And yet there have been few serious prosecutions of the people involved in the Oil-for-Food scandal. On July 13, 2006, South Korea's Tongsun Park—a businessman implicated in a 1970s influence-peddling scandal dubbed Koreagate—became the first person convicted for Oil-for-Food corruption. According to U.S. Attorney Stephen Miller, Park "had arranged to receive millions of dollars in cash to influence top UN officials, including former UN secretary-general Boutros Boutros-Ghali."[28] He is awaiting sentencing.

A few others have been snared: Texas oilmen Oscar Wyatt and David Chalmers were indicted, and Alexander Yakovlev pled guilty to fraud and money laundering. A French businessman has also been charged. Still, the fact remains that almost no one within the UN has been punished for their role in the scandal. Benon Sevan resigned from his UN job under pressure but has not been indicted. Annan remains unanswerable, protected by his diplomatic immunity. No United Nations official has been criminally prosecuted. None of the foreign beneficiaries of Saddam's vouchers has been tried or convicted. In the United States, only the hapless Tongsun Park has been convicted. Fourteen billion dollars stolen from a program designed to feed the poor people of Iraq, and practically no one is being held accountable—not in the U.S. or in Iraq or in the United Nations or in Russia or in France or in Germany or in any of the sixty-six countries where people paid or received bribes.

It is as if the international criminal justice system had simply taken a pass—a profound failure that underscores the near impossibility of prosecuting United Nations scandals. Protected by diplomatic immunity, UN officials cannot be prosecuted; moreover, the difficulty of chasing down a worldwide scandal with so many conflicting jurisdictions and powerful political interests is obvious.

Only the Kofi–Kojo Annan relationship through the Swiss-based Cotecna Inspections has been thoroughly investigated. Kojo Annan is resolute in denying that his work had anything to do with the United Nations or his father's position as its secretary-general, despite his having been employed with Cotecna when the company received its Oil-for-Food contract. As Fox News reported in 2004, however, U.S. congressional investigators have turned up a fax from Cotecna headquarters to Kojo Annan, urging him to become involved with the United Nations. "The Aug. 28, 1998, fax praises Kojo's work at a meeting of world leaders in South Africa and adds: 'Your work and the contacts established at this meeting should ideally be followed up at the September 98 UN General Assembly in New York.'" [29]

Fox also reports that Kojo's phone bills show frequent calls to the UN, and that he registered in a New York hotel as "K Annan, United Nations." [30] Even more telling is that for four years after Kojo Annan left Cotecna, the company paid him $2,500 a month.

Although none of the bribes, payoffs, kickbacks, or other illegal payments made directly to UN officials by the Saddam Hussein regime has led to any indictments or convictions, the organization hastened to clear its secretary-general from charges of conflict of interest. The *Washington Post* reported that the Volcker panel "investigating abuse in the Iraq Oil-for-Food Program said there is no evidence that UN secretary-general Kofi Annan used his influence to steer a multi-million-dollar contract to a Swiss company that employed his son. But the panel's report faulted Annan for inadequately scrutinizing the deal to determine whether his son's involvement posed a conflict of interest, and it accused Annan's senior advisers of misusing Iraqi funds and shredding relevant documents as the investigation began. It also indicated that Annan may have initially misled investigators about contacts he had with senior executives at his son's company before they won a UN contract. [31]

But even as the panel's report headlined the exoneration, it chronicled investigative details that were quite disturbing. It turns out that Kojo Annan may have made considerably more than the $30,000 a year previously reported. The Volcker report said that he "may have earned as much as $485,000 in consulting fees from Cotecna while it conducted millions of dollars in business in Iraq for the United Nations." [32] Equally disturbing is the panel's accusation that Secretary-General Kofi Annan may have misled the investigation when he denied "participating in two meetings with Cotecna's chairman, Elie-Georges Massey, before the contract was awarded."[33] After investigators showed the secretary-general a computer record of the two meetings, "he acknowledged the encounters but said the two never discussed Cotecna's Iraq contract."[34]

Volcker said that "Kojo Annan actively participated in efforts by Cotecna to conceal the true nature of its continuing relationship with him."[35] The panel also disclosed that Iqbal Riza of Pakistan, Annan's former chief of staff, had approved destruction of his backup computer files covering the first three years of the Oil-for-Food Program. His April 22 decision came one day after the Security Council officially endorsed Annan's decision to appoint Volcker. Riza claims that the order was not an attempt to conceal information, and that documents were "simply extra copies" of records that UN officials typically destroy each year. He said that his own policy was to hold on to the files for five years. "In hindsight, I can say, 'Oh, goodness, I shouldn't have.'"[36] His protestations of innocent intent fly in the face of a Fox News report that the shredding of documents took him seven months to complete.[37]

Spinning Volcker's harshly critical report as an "exoneration," the secretary-general called it "a great relief after so many distressing and untrue allegations have been made against me."[38]

The single worst thing about the scandal is how surely it undermines the very concept of the Western alliance and of the diplomatic consultations that lie at its core. When allies like France are getting paid off to back Saddam Hussein, how is collective action possible? Indeed, the very ambassador that France sent to the United Nations for much of the Oil-for-Food period, Jean-Bernard Merimee, has since been implicated in receiving oil bribes and is currently under investigation.

After such an affair, how can the United States feel comfortable trusting the advice of its continental allies? Are they speaking from knowledge and wisdom or motivated by bribery and greed? We will always have to wonder.

SEXUAL ABUSE BY UNITED NATIONS PERSONNEL

History's sad pages are filled with lurid tales of sexual abuse, rape, and violation of women by invading forces. But it was particularly shocking to read that United Nations *peacekeeping* forces, assigned to trouble spots around the world, have been guilty of the same kind of conduct.

The United Nations deploys ninety thousand peacekeeping troops in fifteen locations throughout the world. There are currently peacekeeping operations in the Western Sahara, the Democratic Republic of the Congo, Ethiopia and Eritrea, the Ivory Coast, Liberia, Burundi, Sudan, Haiti, India and Pakistan, Kosovo, Georgia, Cyprus, Lebanon, the Middle East, and the Golan Heights.[39] Most of those who serve in these troubled areas are honorable men and women furthering the cause of world peace. But some are guilty of a sordid array of sexual offenses.

In all, there have been 221 investigations of sexual improprieties—including pedophilia, prostitution, and rape—by United Nations civilian employees and military peacekeepers from February 2003 to October 2005.[40]

The *Times* of London, in an article headlined "The UN's Abu Ghraib," details some of the more lurid cases: [41]

- In December 2004, a French UN logistics expert in the Democratic Republic of the Congo was arrested for making pornographic videos with young girls. When police arrested the man at his home, they found a twelve-year-old girl he was allegedly about to rape as well as some fifty pornographic photographs.

- In May 2005, four Nepalese soldiers flew home from the Congo to face charges of sexual abuse stemming from 2003.

- According to the *Times*, two Russian UN peacekeeping pilots based in Mbandaka, a city in western Congo, allegedly taped sex sessions with minors, paying them with jars of mayonnaise and jam. The accused men have since fled to Russia.

- For a year, a Moroccan peacekeeping unit in Kisangani, a city in the northern part of Congo, allegedly hid in its barracks a soldier accused of rape. Moreover, two UN officials—one from Canada and the other from Ukraine—left the Congo after it was alleged that they had impregnated local women.

- There are sixty-eight allegations of sexual abuse involving UN personnel in the northeast Congo town of Bunia, where UN peacekeepers have been stationed since May 2003. According to Joseph Loconte, in early 2005, the United Nations Children's Fund (popularly known as UNICEF) treated some two thousand victims of sexual violence there over a period of just several months.

Recently, a thirty-four-page internal United Nations report, obtained by the *Washington Post,* accused UN peacekeepers from Morocco, Pakistan, and Nepal of "seeking to obstruct UN efforts to investigate a sexual abuse scandal that has damaged the United Nations' standing in Congo."[42]

The report cites sixty-eight cases of "alleged rape, prostitution and pedophilia by UN peacekeepers from Pakistan, Uruguay, Morocco, Tunisia, South Africa, and Nepal. UN officials say they have uncovered more than 150 allegations of sexual misconduct throughout the country as part of a widening investigation into sexual abuse by UN personnel that has plagued the United Nations' largest peacekeeping mission, UN officials said."[43]

"Sexual exploitation and abuse, particularly prostitution of minors, is widespread and long-standing," says a draft of the internal July report, which has not previously been made public. "Moreover, all of the major contingents appear to be implicated."[44]

The UN peacekeeping operation in the Congo seems particularly out of control. Its chief was recently forced to resign after he was alleged to have consorted with prostitutes at a local bar.[45]

While UN officials decry the situation, the organization continues its policy of handing out free condoms to its soldiers as part of its AIDS prevention work.

The UN's under-secretary-general for Peacekeeping Operations, Jean-Marie Guehenno, told the *Times* of London, "The fact that these things happened is a blot on us. It's awful. What is important is to get to the bottom of it and fight it and make sure that people who do that pay for what they have done."[46] Then-secretary-general Kofi Annan echoed his concern: "I am afraid there is clear evidence that acts of gross misconduct have taken place," he admitted. "This is a shameful thing for the United Nations to have to say, and I am absolutely outraged by it."[47]

"I have long made it clear that my attitude to sexual exploitation and abuse is one of zero tolerance, without exception, and I am determined to implement this policy in the most transparent manner," Annan said.

But, as the *Times* reports, Jordan's Prince Zeid Raad Al Hussein, a special adviser to Annan who led one investigative team, said, "The situation appears to be one of 'zero-compliance with zero-tolerance' throughout the mission."[48]

United Nations peacekeeping forces are provided by individual countries to serve under UN command. When cases of sexual abuse are reported, the soldiers are sent home to their respective countries for discipline or prosecution. In some countries the disciplinary process is severe. In others, it is more like a slap on the wrist. The fact that the offense cannot be punished by the United Nations, even though it is committed by UN personnel on UN duty, is the military equivalent of the inability to prosecute the corrupt officials in the Oil-for-Food scandal.

Besides, as Sarah Martin of Refugees International writes, "If a soldier is found guilty [of abuse], that person is sent back to his country for discipline. It is very difficult, if not impossible, for victims and their families to determine what, if any, actions have been taken."[49]

If the Oil-for-Food scandal undermines confidence in the diplomatic consultations that underlie United Nations action, and the administrative framework that implements its decisions, the sexual abuse scandals cast doubt on the integrity and legitimacy of UN peacekeepers. The scandal underscores the question: Who will keep the peace among the peacekeepers?

RWANDA: HOW THE SECRETARY-GENERAL MADE THE GENOCIDE POSSIBLE

It is bad enough that the United Nations—and the United States under Bill Clinton—stood idly by while hundreds of thousands of Rwandans were massacred—many hacked to death by the rival Hutus—but the weapons the Hutus used came, in part, through the arms dealings of the Egyptian foreign minister Boutros Boutros-Ghali, who went on to become the secretary-general of the United Nations. Boutros-Ghali served while the Rwandan massacres were taking place using the very weapons he sold to the Hutus—and he did nothing to stop them.

According to *The Observer,* a British newspaper, Boutros-Ghali "played a leading role in supplying weapons to the Hutu regime, which carried out a campaign of genocide against the Tutsis in Rwanda in 1994. As minister of foreign affairs in Egypt, Boutros-Ghali facilitated an arms deal in 1990, which was to result in $26 million (£18m) of mortar bombs, rocket launchers, grenades, and ammunition being flown from Cairo to Rwanda. The arms were used by Hutus in attacks which led to up to a million deaths." [50]

Boutros-Ghali had lots of help. French banks "helped launder the international aid money that was used to pay for that [arms shipment from Boutros-Ghali] and other arms shipments." Particularly gruesome was the importing of 580,000 machetes from China; the Hutus used the weapons to hack the Rwandans to death. [51]

In her book *A People Betrayed: The Role of the West in Rwanda's Genocide,* Linda Melvern writes that in 1990 Boutros-Ghali approved an initial arms deal with the Hutus for $5.8 million, which led to larger arms purchases further down the line. As foreign minister for Egypt, Boutros-Ghali has said that selling arms was part of his job.

The Observer notes that "the deal was never disclosed: the weapons were smuggled into Rwanda and disguised as relief material. At the time there was an international outcry at human rights abuses by the Hutu government as thousands of Tutsi were massacred." [52]

"Asked about the wisdom of an arms deal at such a sensitive time, Boutros-Ghali said he did not think that a 'few thousand guns would have changed the situation'. His contacts with the Hutu regime have never been investigated." [53]

The United Nations' and the United States' response to the horrible genocide in Rwanda was mitigated by their bad experience in Somalia just two years before. There, UN forces, including a contingent of twenty-five thousand American troops, were sent to distribute food in order to prevent imminent starvation. But the United States and the UN soon became embroiled in the local conflicts among warlords. When eighteen U.S. soldiers were killed and eighty-four wounded in a firefight with local warlords on October 3–4, 1993, President Clinton ordered a gradual withdrawal of American forces.

Burned by this bad experience on the African continent, U.S. and UN officials were reluctant to intervene in Rwanda, even though Lt. General

Romeo Dallaire, commander of the UN forces in Rwanda, said that only five thousand more soldiers would be needed to stop the killings. The Security Council at the UN rejected Dallaire's plan and ordered him not to intervene. When ten Belgian peacekeeping soldiers were killed, Belgium withdrew its 2,500 troops, leaving a force of 5,000 mainly African soldiers who did little to stop the slaughter of at least 300,000 Rwandans.[54]

In a 2004 interview, Boutros-Ghali told PBS that Rwanda was "one of my greatest failures. I failed in Rwanda. My failure in Rwanda is greater than my failure in Somalia, because in Somalia, I was aware of what will happen once the international community withdrew from Somalia. But I was not aware of the degree of disaster in Rwanda, so it is a double failure. It is a failure that I was not able to convince the members of the Security Council to intervene, and it is another failure that I was not able to understand from the beginning the importance of what was going on. . . . It took me weeks before suddenly we discovered that it was genocide. So this is another kind of failure."[55]

LARCENY AT THE UN

While the United Nations focuses on wars throughout the world and wrestles with weighty decisions about international economic sanctions, military action, and the like, it apparently pays little attention to overseeing its own employees. With a management structure that has not been updated as the organization has expanded, the UN has become a target for bribery and chicanery well beyond the Oil-for-Food scandal. The epicenter of this fraud is the procurement department, which reports to the secretary-general.

Everything the UN buys, from "cappuccino and paper clips at UN headquarters" to "air freight services and food rations for peacekeeping troops worldwide," as Claudia Rosett and George Russell wrote in an investigation for Fox News, goes through the secretary-general's procurement office. In all, the office controls $1.4 billion in purchases annually, with about a quarter of the funds coming from American taxpayers.

But on August 8, 2005, Alexander Yakovlev, the head of the office, "pleaded guilty to federal charges of corruption, wire fraud, and money laundering." Rosett and Russell reported that "federal investigators have now alleged that from 2000 on, Yakovlev . . . [was] transferring bribe money to the head of the United Nations' own budget oversight committee, a Russian named Vladimir Kuznetsov[,] via the Antigua Overseas Bank in the West Indies. Allegedly the bribe money was in exchange for providing inside information to companies seeking UN contracts."[56]

Yakovlev was no bit player. He "held various portfolios in procurement during his twenty-year career [at the UN] in places as far flung as Africa, Asia, and the Middle East. He even managed the architectural contract for the UN's proposed $1.2 billion renovation of its Manhattan headquarters," a project that is running way over budget.

Rosett and Russell reported that "the amounts at issue in this alleged Russian bribery ring are quite likely far larger than the 'several hundreds of thousands' so far cited by federal investigators." In his report, Paul Volcker "noted in passing that Yakovlev had received more than $950,000 in bribes from companies that 'collectively won more than $79 million in United Nations contracts and purchase orders.' "[57]

But the bribery and fraud extend far beyond the confines of UN headquarters. In 2005, the UN disclosed that the commanding officer of its Ukrainian peacekeeping contingent in Lebanon had been engaged "in significant financial misconduct."[58]

Stung by such accusations, the UN commissioned a study of its procurement department by the U.S.-based firm Deloitte Consulting. The results were devastating. The report said that ethics and integrity are "not sufficiently supported by management and is not a conspicuous element of the [department's] culture."[59]

As the Fox News report points out, the study concluded that "UN procurement employees themselves are the only control on the department and that 'significant reliance on people leaves the UN extremely vulnerable to potential fraudulent or corrupt activity and [with] limited means to either prevent or detect such actions.' "[60]

Deloitte said, "There does not appear to be a single, recognized and well understood Code of Conduct governing ethics and integrity expectations and requirements for [department] employees."[61]

The United Nations even has a hard time renovating its own headquarters. The cost of the projected renovation has soared from an estimated

$1.2 billion to over $1.7 billion. The U.S. ambassador to the United Nations, John Bolton, said that the lax supervision of the project constitutes "an open invitation to those who may seek to defraud or abuse the system."[62]

As if the scandals in the procurement office were not enough, on July 27, 2006, the U.S. attorney and the FBI broke up a drug ring that was operating *out of the United Nations mail room,* using diplomatic pouches to smuggle twenty five tons of illegal drugs into New York during the past eighteen months. Khat—an illegal stimulant grown in the Horn of Africa—was the principal drug the ring brought into the country. "Chewing the leaves has long been a custom in the countries of [East Africa]. Immigrants have brought the habit to America, where the active chemical in the leaf is as illegal as heroin."[63]

The employee at the center of the scandal, the imaginatively named Osman Osman, had worked for the UN for twenty-nine years; the government called him "an important cog in the largest khat trafficking enterprise America has known."[64] Forty-four other defendants were named in the indictment, and fourteen were still at large when the bust was announced.

The government said that revenues from the sale of khat went to Somali warlords, some with al Qaeda connections, who dominate Somali politics. The street value of the illegal drugs was said to exceed $10 million.

All of the procurement scandals, and the mail-room drug bust, have followed from investigations by the U.S. Attorney's Office for the Southern District of New York, which has jurisdiction over crimes committed at the UN headquarters on Manhattan's East Side. But the fact that the United States had to prosecute United Nations corruption, simply because it happens to be where the organization is located, illustrates the UN's woeful failure to police itself.

REFORMING THE UN MESS

George W. Bush's temporary ambassador to the UN, John Bolton, made a priority of controlling the mismanagement and cost overruns that plague the international body. But he had only limited success before leaving the office at the end of 2006. He ran up against a wall of opposition from small nations eager to water down his proposals for reform and accountability. "To be frank," Bolton told the U.S. Senate Foreign Relations Committee, "the overall results [of his reform efforts] have not been particularly encouraging. There has been some movement, but no notable successes so far."[65]

Bolton is up against what Paul Volcker, after investigating the Oil-for-Food scandal, called "the culture of inaction" that pervades the UN when it comes to reform. Bolton said "changing that culture and adapting it to modern-day management and accounting norms is no small task, but failure to do so is simply to invite future scandals."[66]

But Bolton did succeed in pushing through a $950 million annual spending cap on the UN budget, saying that the scandals at the UN "would have [made it] irresponsible for member states to approve a 'business-as-usual two-year budget." He said that "by securing passage of a limit on UN spending through imposition of [a] spending cap on the two-year budget, the United States scored a significant diplomatic victory in a consensual manner, despite many member states' initial shock at the suggestion of using the budget as a lever to secure further progress on reform efforts."[67]

The UN's own inspector general, the Office of Internal Oversight Services (OIOS), is charged with auditing, investigating, and evaluating all UN activities under the secretary-general. But Bolton complains that the OIOS is "beholden to those it is responsible for investigating," particularly for its financial support. David M. Walker, the comptroller general of the United States, said that "UN funding arrangements constrain OIOS's ability to operate independently as mandated by the General Assembly and required by international auditing standards OIOS has adopted. . . . OIOS depends on the resources of the funds, programs, and other entities it audits. The managers of these programs can deny OIOS permission to perform work or not pay OIOS for services. UN entities could thus avoid OIOS audits and investigations, and high-risk areas can be and have been excluded from timely examination."[68]

Bolton has rightly called for an end to this peculiar system, under which those suspected of fraud must give investigative permission to the cop who might catch them—while controlling funding for any such inquiry. Though his appointment was not confirmed by the Senate, after a year at the UN, many of Bolton's critics—including the *New York Times*—praised his performance, surely due in part to his work to curb UN corruption.

One of the UN success stories Bolton celebrated was UNICEF, the United Nations' Children's Fund. "One of the keys to UNICEF's success," he observed, "is its emphasis on measurable results, which document and prove to existing and potential contributors that their money is being well-spent"[69] The critical difference between UNICEF and other UN programs is that the Children's Fund gets most of its money through voluntary contributions from governments and individuals throughout the world. Remember "Trick-or-treat-for-UNICEF"? Catherine Bertini, former UN under-secretary-general for management and former head of the World Food Program (WFP), another group that relies on voluntary funding, says that the system "creates an entirely different atmosphere at WFP than at the UN. At WFP, every staff member knows that we have to be as efficient, accountable, transparent, and results-oriented as is possible. If we are not, donor governments can take their funding elsewhere in a very competitive world among UN agencies, NGOs, and bilateral governments."[70]

Part of the problem at the United Nations, as at any governmental body, is the accumulation of old programs that have outlived their reason for being and simply clutter up the budget and suck up funds to pay their bureaucrats year after year. Bolton proposed a mandatory review, after five years, of any UN-mandated program, to subject it to scrutiny and determine if it is still needed. But the seventy-seven nations that pay only 12 percent of the budget at the UN opposed the reform. They pressed to exclude three-quarters of the mandates from the review. Bolton said "these countries have made clear their interest in the status quo on this issue, which has resulted in active opposition to any genuine reform."[71]

HUMAN RIGHTS: MAKING THE VIOLATORS THE ENFORCERS

The United Nations was created at the San Francisco Conference in 1946. That same year, in light of the revelations of the Holocaust, the UN Commission on Human Rights came into being. Eleanor Roosevelt led the effort, which produced the Universal Declaration of Human Rights, which the General Assembly adopted in 1948. As Ryan Balis of the National Center for Public Policy Research notes, "the Declaration's thirty articles delineated such fundamental rights as social and political freedom, equal protection under the law, freedom of speech and assembly, and the right to own property. The Declaration also prohibits slavery, torture, discrimination, and arbitrary arrest, detention, or exile."[72]

But the record of the Human Rights Commission in enforcing the Declaration has been pitiful. Joseph Loconte, the Heritage Foundation's William E. Simon Fellow in Religion and a Free Society, explains why: "It is a regrettable, yet widely recognized fact: repressive governments seek membership on the [Human Rights] Commission to escape scrutiny and censure."[73]

In 2003, the commission was chaired by none other than Libya—one of the most repressive governments in the world. Its other members included Iraq, Nigeria, Saudi Arabia, and Cuba, notorious human rights violators. Even the Sudan, now perpetrating the Darfur genocide, became a member. As President Bush said in a speech to the UN, "When this great institution's member states choose notorious abusers of human rights to sit on the UN Human Rights Commission, they discredit a noble effort and undermine the credibility of the whole organization."[74]

In his testimony before the House subcommittee on Global Human Rights and International Operations, Loconte recounted how in 2003 the Commission on Human Rights voted to remove Sudan from a list of countries requiring special monitoring. "The Commission reached the nadir of its corruption last year, however, when the Sudanese government—repeatedly accused of gross human rights abuses in Darfur—was reelected as a Commission member in good standing."[75]

In 2001, the General Assembly had the audacity to vote the United States off the Human Rights Commission—on the same day that Pakistan, Sierra Leone, and Sudan were voted on! Amnesty International, no friend of the Bush administration, was quick to condemn the removal of the United States from the commission. It called the action "part of an effort by nations that routinely violate human rights to escape scrutiny."[76] Amnesty accused members of the commission of failing to do their job, succumbing instead to political and economic pressures. "The U.S.," the organization said, "was among the few nations willing to actively push for condemnation at the UNHRC of the brutal human rights violations committed by nations like China."[77]

Madeleine K. Albright, who served as United States ambassador to the United Nations before becoming secretary of state, called the rejection "beyond belief." But some liberals were quick to blame the United States for the vote against it. Congresswoman Nita M. Lowey (D-NY), cochair of the

United Nations working group in Congress, said that it was a lack of U.S. "commitment to human rights" that contributed to the rejection. She said that commitment "has fallen victim to the [Bush] administration's laissez-faire attitude toward diplomacy and foreign policy."[78]

The Commission has amassed a remarkable record of paying no attention to human rights abuses. In 1998, Moroccan Halima Warzazi and Cuban Miguel Alfonso Martinez joined to try to kill a UN resolution condemning Saddam Hussein for using "banned chemical weapons" in his war against the Kurds. Saddam had provoked the resolution by dropping nerve gas on five thousand Kurds, killing them. Warzazi was rewarded for his service to human rights by being elected chairperson of the UN Human Rights Commission; Martinez still sits on it as a member.[79] By the same token, Iran has never been criticized by the Commission—despite its record of persecuting those who do not follow its fundamentalist brand of Islam or accept its discrimination against women.

Nor has Zimbabwe, with its horrible massacres and human rights violations, ever been criticized by the Commission. Even Russia has never been cited for its massive and cruel human rights abuses in Chechnya, or its increasing repression at home.

When the United States proposed a resolution criticizing China for its conduct in Tibet and Sinkiang, the "Chinese ambassador choked with indignation" and "not one of the fifty-three member countries" voted for the U.S. proposal.[80] By killing the U.S. resolution, China succeeded in precluding any debate about its human rights record, as it has done successfully every year except in 1995. Xiao Qiang, executive director of the American organization Human Rights in China, said, "With this vote, every human rights victim in China, from tortured Falun Gong practitioners to those imprisoned for advocating democracy, has been turned away at the door, which among all others, should be open to them."[81]

Deputy Assistant Secretary of State for International Organization Affairs Mark P. Lagon described how China and other human rights violators get away with killing efforts to denounce them in the Human Rights Commission. He noted that "some of the world's most egregious violators of human rights work though their regional blocs to gain nomination and election to the CHR in order to protect themselves from criticism. At the same time, there has been a disturbing trend for developing countries to turn away from country-specific resolutions that single out and scrutinize those countries with the worst human rights records. Non-democratic states like Cuba and China have led a vocal campaign to steer the UN away from passing resolutions that condemn serious human rights violations. . . . As a result, these countries, often joined by like-minded nations with similar records of human rights abuse such as Syria, Zimbabwe, and Sudan, have criticized country-specific resolutions as 'selective' and 'politicized.'" [82]

Cuba, for its part, was reelected to the Human Rights Commission "days after Fidel Castro arrested nearly eighty human rights and pro-democracy activists and summarily executed three men for attempting to hijack a ferryboat and escape to Florida." [83]

One proposal on which the Commission did act was to adopt, by a vote of 30 to 20, a resolution calling for the abolition of the death penalty! (Castro must have paid no attention.)

The Geneva-based organization UN Watch has compared the membership on the Human Rights Commission with the ratings compiled by Freedom House, a New York group that monitors human rights. Freedom House rates a nation's human rights record on a scale from 1.1 (best) to 7.7 (worst). As the New York Times reported, "five members of the commission—Cuba, Libya, Saudi Arabia, Syria, and Vietnam—had 7.7 ratings, the worst possible." [84]

But there is one subject on which the Human Rights Commission has been quite outspoken: Israel. Spurred by its Islamic members, 30 percent of the Human Rights Commission resolutions criticizing specific countries have been directed against Israel, according to human rights scholar Anne Bayefsky. [85]

In 2006, the UN members, led by Secretary-General Kofi Annan, finally recognized a need to reform the Commission. But it replaced it with a new membership that is just as bad. Secretary of State Condoleezza Rice's attempt to require a two-thirds vote of the General Assembly to serve on the Council was rejected in favor of a simple majority, even though Secretary-General Kofi Annan had endorsed Rice's proposal. Then a coalition of human rights abusers and their sympathizers hijacked the Council, just as they had done with the discredited Commission.

To show its discontent with the "reform," the U.S. boycotted the Council and refused to seek a seat among its members. Backed by what the Heritage Foundation called "tyrannical regimes such as Burma, Syria, Libya, Sudan, and Zimbabwe," along with North Korea, the Council was born. Its first election saw the elevation to membership of such repressive countries as Algeria, Cuba, China, Pakistan, Russia, and Saudi Arabia. In its first emergency session, it condemned Israel, in a one-sided resolution that ignored the Palestinian suicide bombings and rocket attacks.[86]

Human Rights Watch was quick to criticize the biased action. It said that "by adopting a politicized resolution that looks only at Israeli abuses in the current conflict, the Human Rights Council undermined its credibility and wasted an opportunity to protect civilians in the region. . . . The council decided to establish a commission of experts to investigate deadly attacks by Israel, but took no action with regard to Hezbollah's murderous abuses."

As Peggy Hicks of Human Rights Watch observed, "The one-sided approach taken by the Human Rights Council is a blow to its credibility and an abdication of its responsibility to protect human rights for all." Hicks, the global advocacy director at Human Rights Watch, says wryly, "This is a poor way to launch a new institution."[87] John Bolton had it right when he warned that the new Human Rights Council "could . . . replicate the same flaws of old" and fail to do much of anything to protect human rights.[88] The fact that Russia, China, Cuba, and Algeria, serial human rights abusers all, will be the ones sitting in judgment on the Council makes it more than clear that nothing will be accomplished.

In one important respect, however, Bush and Bolton met with some success: the establishment of a UN Democracy Fund "to provide grants to civil society, governments, and international organizations in order to carry out programs that strengthen democracy." Deputy Secretary Lagon reports that fifteen nations have donated $43 million to the fund.

But the abuses of the Human Rights Council go on—striking down to the very core of the United Nations' mission. Originally intended to promote world peace and human rights, the organization now devotes much of its time to obliterating the latter, leaving us with only the former. But peace without liberty is not peace at all. It is repression. The actions of the Human Rights Council force us to hark back to Patrick Henry's age-old question: "Is life so dear and peace so sweet as to be purchased at the price of chains and slavery?"[89]

ACTION AGENDA

Ambassador Bolton's reforms, designed to curb financial and other abuses at the United Nations, were an excellent start, but the question remains: how is the United Nations going to deter fraud in the future? How can the nations of the world trust the UN to be the impartial administrator and implementer of their decisions if programs such as Oil-for-Food are filled with massive fraud, graft, and corruption and soldiers sent abroad as peacekeepers are able to wreak mayhem in their wake?

This, in short, is what happens to a billion-dollar organization when it has no judicial or prosecutorial arm. Unless the transgressions happen to run afoul of the U.S. Attorneys' Office where the UN headquarters is located, there is no way to punish or discipline UN employees who are guilty of wrongdoing. There is no prosecutor. There are no courts. There is no jail. There is no criminal code. There is no jury. There is no judge. There is no courtroom.

Instead, the only punishment for a transgressing employee is to get fired—small disincentive against the massive opportunities for fraud and abuse invited by an operation with as much money, and as little oversight, as the United Nations.

The United States is by far the largest contributor to the UN—we pay far more than our share, while many nations pay far less. The sidebar shows a list of donors in order of their annual contributions.

THE UN BUDGET:

THE 15 LARGEST DONORS[90]

United States	$423 million
Japan	$332
Germany	$143
United Kingdom	$105
France	$103
Italy	$83

(continued)

Canada	$48
Spain	$43
China	$35
Mexico	$32
South Korea	$31
Netherlands	$29
Australia	$27
Brazil	$26
Switzerland	$20

It's hard to understand why China, with an economy approaching the world's third largest, pays only $35 million while countries like Canada, Spain, and Italy, whose economies are much smaller, pay more—especially in view of the veto power China wields on the Security Council.

The U.S. contribution to the UN does come with some benefits. For one thing, it gives us tremendous leverage to demand reform. Bolton's calls for action were historic, but if they are going to take hold they need broad and intense support throughout the United States.

But all of this raises a more fundamental question: Why are we in the United Nations in the first place? Clearly some sort of international organization is needed—a place where the world can speak and make its collective will manifest. But why this flawed organization, which puts its power into the hands of governments run by dictators and their compliant military backers?

In the World Trade Organization (WTO), no nation is allowed in until it can demonstrate that it has a free, open market economy. Russia, for all its global power and economic might, is kept out of the WTO because it fails to meet the required criteria. Why shouldn't an organization with similar admission standards spring up to replace the UN?

For that matter, why not restrict membership in future international organizations to democracies?

The United Nations was a phrase first given to the allies in World War II. With the notable exception of the Soviet Union, it was considered to be a coalition of democracies. There was a qualitative standard for membership.

At the UN's inception, neither Japan nor Germany were admitted as members; only *after* they demonstrated that they were embracing democratic norms were they admitted.

Why should the votes of the U.S., which represents the opinions of 300 million people, be equal to that of China, which represents only the small segment which runs the country?

Some argue that the UN must include bad guys as well as good guys if it is ever to discipline the former effectively. But the UN has shown little ability to discipline misbehaving states—particularly the big ones that sit on its Security Council as permanent, veto-wielding members. The moral effect of including only democracies in the United Nations, and excluding countries whose sham elections do not measure up to this standard, would be huge. The excluded countries would be under heavy pressure to alter their means of governance. It would be a decision, in effect, by the democracies of the world to limit the extent to which they are willing to deal with nondemocratic nations.

A United Nations made up only of democracies would carry the tremendous weight and moral authority that the UN now lacks. It would truly speak for the people of the world, not for a handful of tyrants who enforce their rule by military means and police-state tactics. Nations like Cuba, China, Algeria, Syria, Burma, and Saudi Arabia, to name a few, do not belong in an international organization—and they will not until they adopt democracy as their true form of government.

HYPOCRISY AT THE ACLU:
*A*gainst *C*riticizing the *L*iberties *U*nion

Who would ever have thought that the American Civil Liberties Union would attempt to muzzle the free speech of its own board members?

That's exactly what it tried to do.

In early 2006, a memo was circulated to the members of the ACLU Board of Directors with a proposal to shut up any board members who might be tempted to publicly criticize the ACLU staff or another board member. According to the *New York Times,* the proposed policy stated that

> a director may publicly disagree with an ACLU policy position, but may not criticize the ACLU board or staff . . . where an individual director disagrees with a board position on matters of civil liberties policy, the director should refrain from publicly highlighting the fact of such disagreement. . . . Directors should remember that there is always a material prospect that public airing of the disagreement will affect the ACLU adversely in terms of public support and fund-raising.[1]

Hard to believe, isn't it?

Many supporters of the ACLU were horrified at what appeared to be nothing more than a heavy-handed move to stop any dissent in the group. Is the ACLU trying to prohibit a public display of disagreement or rebellion? What's wrong with this picture? Isn't the ACLU—the undisputed champion of individual freedom—supposed to be the one criticizing the government about suppressing opposition? And now the group actually wants to stop its own board members from publicly criticizing them?

The ACLU is all for freedom of speech—as long as no one's talking about them.

Famed First Amendment defender and former ACLU board member Nat Hentoff criticized the proposal. "For the national board to consider promulgating a gag order on its members—I can't think of anything more contrary to the reason the ACLU exists." [2] He quipped: "You sure that didn't come out of Dick Cheney's office?" [3]

The ACLU exists, in large part, to protect freedom of speech. According to its website, "The mission of the ACLU is to preserve all of these protections and guarantees:

> Your First Amendment rights—freedom of speech, association and assembly. Freedom of the press, and freedom of religion supported by the strict separation of church and state." [4]

So what led the nation's staunchest defender of free speech to try to silence its own board members?

Apparently some board members had actually dared to alert the press to certain policies of the organization that they felt were alien to its core purpose and, in fact, downright hypocritical. And somebody up there didn't like it. Based on published reports, what ensued was the equivalent of a boardroom catfight.

It all started in 2004 when two national board members, Michael Meyers (not the comedian, but a founder of the New York City Civil Rights Coalition) and Wendy Kaminer (a lawyer and author) publicly denounced the ACLU and its executive director, Anthony Romero. Their grievance? That Romero had certified to the federal government that the organization would not knowingly hire anyone whose name appeared on a watch list of suspected terrorists created by the U.S. government, the European Union, and the UN. Signing the certification was a condition for qualifying for $470,000 of government-collected charitable donations from federal workers.

The board members, who were not told about the ACLU's certification until months after it was executed, were concerned that the ACLU had contradicted its own policy on "no-fly" lists. The ACLU had actually filed a suit seeking to prohibit the federal government from using such lists and claiming that the use of the list violated constitutional rights in some cases. In the heated discussion that followed, some board members argued that the agreement to check the no-fly list against the names of its own employees and applicants was tantamount to supporting what they said was the government's reprehensible policy.

Romero had it all figured out, though. He assured the board that although he had agreed to and signed the required certification, he had also deliberately avoided reading the no-fly watch list. That way he would have no way of knowing whether anyone he hired appeared on the list. No problem. "We would never terminate or kick off board members or staff members because of their associational rights." [5]

Nadine Strossen, ACLU's president at the time, commended Mr. Romero for his "clever" interpretation of the pledge.[6]

"Do we do more harm than good by spurning money by certifying something that is plausible but not the only plausible interpretation?" she asked. "It's completely a debate about strategy, not principle. I think Anthony handled it completely appropriately."[7]

A motion to retract the signed certification failed to pass at the board meeting. But shortly afterward, following an article about the issue in the *New York Times,* the Executive Committee agreed to send the money back.

Some board members were unhappy that Romero had also accepted a $68,000 grant from the Ford Foundation that contained language based on the Patriot Act that was intended to make sure that grant money was not used to finance terrorist activities. The ACLU has aggressively fought the Patriot Act, and Romero himself has been one of its harshest critics. What made the Ford Foundation issue even worse for some board members was that Romero had apparently been involved earlier in advising his former employer, the Ford Foundation, to follow the dictates of the Patriot Act.

Eventually, the board sent the money back to the Ford Foundation.

But that didn't solve all of the problems between Romero and the board.

Meyers and Kaminer continued their public criticisms of ACLU actions they believed were at odds with the values and purposes of the organization. When they learned that the ACLU was involved in data mining to identify prospective members and donors, all hell broke loose.

Apparently, the ACLU had hired Grenzebach Glier & Associates to identify prospective donors, in part, by using its software to scour databases such as those listing donors in federal elections, Dun & Bradstreet, Lexis-Nexis, and other sources. This was a big change in the fund-raising strategy: in the past the organization had identified donor zip codes and checked SEC filings, but privacy concerns had kept them from going further.

Meyers insisted that the new data-mining project, begun without board approval, violated the ACLU's own privacy policy, which assured members that information about them would remain confidential. "If I give the ACLU twenty dollars," he pointed out, "I have not given them permission to investigate my partners, who I'm married to, what they do, what my real estate holdings are, what my wealth is, and who else I give my money to."[8]

Wendy Kaminer agreed: "It goes against ACLU values to engage in data mining on people without informing them. It's not illegal, but it is a violation of our values. It is hypocrisy."[9]

The ACLU has been a frequent critic of government and businesses for violating individual privacy rights. In fact, the ACLU's own website specifically addresses the issue:

> For centuries, bankers used to pride themselves on being discreet and confidential about their customers' business. But today that tradition of discretion and confidentiality is breaking down, and consumers' financial privacy is in a sorry state indeed.
>
> *The problem lies not just with banks, but also with insurance companies and many other corporations who gather details about the financial lives of Americans, and increasingly see those details as a valuable resource to be mined for profit.*[10] [emphasis added]

Sounds like exactly what the ACLU was doing, doesn't it? Meyers pointed out that the ACLU privacy policy, which was posted for public consumption on the website, assured members that it would not gather information about them without their permission. After Meyers's complaints, that language was quickly and quietly deleted. Now the privacy policy discloses that donor information is shared with other foundations and gives members an opportunity to opt out on having their data shared.

Meyers learned about the data-mining initiative only by accident. But after former New York attorney general (now governor) Eliot Spitzer's office informed the ACLU that his office was investigating whether it had violated its own privacy policy, the Executive Committee scheduled a meeting to discuss the matter. Meyers exercised his right as a board member to listen to the meeting by phone.

Guess what the first decision voted at that meeting was? To exclude Meyers from hearing the deliberations.

That's what the ACLU does when there's dissent.

Earlier that year, Romero had signed a consent decree with Spitzer's office regarding the security of the ACLU website. A consent decree is a legal agreement between a party and a government agency in which the party agrees to take specific actions or to refrain from taking specific actions. It is usually enforceable by a court. The consent decree Romero signed required him to notify the ACLU board of the decree within thirty days.

Romero waited almost six months to tell the board. The Stanford- and Princeton-educated lawyer claimed that he hadn't read the consent decree carefully (just like he wasn't going to read the no-fly list), and therefore didn't realize it included a disclosure requirement. Most organizations that were forced into a consent decree with a state attorney general—particularly one named Eliot Spitzer—would read it carefully and then follow it to the letter.

After Meyers and Kaminer's statements to the press, there was a move to throw them off the ACLU board. The resolution failed, but descriptions of the antics at the meeting suggest the conduct of a third grade disciplinary committee rather than the board of a $100-million-a-year national advocacy organization.

To put it bluntly, Romero had a hissy fit. By his own admission, he lost his temper and flew into a fury, castigating Kaminer for speaking to the press about her disagreement with another ACLU position. Other board members described his history of bizarre behavior in detail. Evidently, it was his habit to call individual board members out of the meeting for a scolding when he felt they weren't supporting him—even if they never said a word. One colleague, Allison Steiner, was ordered out into the hallway, "where he chastised her for the look on her face when he was criticizing Ms. Kaminer." [11] (This is not a joke.)

Steiner recounted the episode in an e-mail message to the board that was published in the *New York Times:* "Anthony went on to say that because I was Wendy's 'friend' and did not appear ready to join him in 'getting rid of her' (by, among other things, lobbying her affiliate to remove her as its representative), I was no better than she was, and then stormed off angrily." [12]

Another board member, David F. Kennison, was also called out to the corridor woodshed. His transgression: he had openly apologized for not supporting Ms. Kaminer when Romero savaged her.

In an e-mail message, Kennison later indicated that Romero "told me that he would 'never' apologize to the target of his outburst and that his evaluation of her performance as a member of this board was justified by information he had been accumulating in a 'thick file on her.'" [13] Kennison asked Romero if he was going to start a file on him. According to Kennison, "He asked me what made me think that he didn't already have a file on me." Romero claimed that Kennison had "provoked" him, and he denied that he had a file on Kaminer.

Keeping files on ACLU Board members? Romero sounds more like J. Edgar Hoover than an ACLU official.

After the unsuccessful attempt to dump Meyers and Kaminer, a committee was formed to address the issue of press criticism. That's where the gag rule proposal came from. Many people believe it would have passed easily, but for a story that appeared in the *New York Times* before the board meeting where it was scheduled for consideration. When asked for comment on the proposal, Romero told the *Times*—can you guess?—that *he hadn't had a chance to read it.*

A day before the meeting, a lawyer from the New York attorney general's office called the ACLU and indicated that his office might well have a problem with a public charity prohibiting speech by its members. [14] The Executive Committee discussed the call from the AG's office, but decided not to tell the board about the call.

You see, the board had invited a *New York Times* reporter to its meeting to demonstrate its openness in the face of criticism in the newspaper. It didn't want to close the door in the reporter's face and go into executive session. It certainly didn't want to let the *Times* know that the Attorney General had called them to account for suppressing free speech. So the ACLU—supposed champions of liberty—decided to shut up and not tell the board about the attorney general's call.

With the *Times* reporter in attendance and the attorney general breathing down its neck, the ACLU leadership wisely decided that the best course would be to move on and not to vote on the gag rule at that particular meeting.

Ultimately, Nadine Strossen announced that the ACLU would drop all efforts to keep board members quiet, and would not interfere with their First Amendment right to speak out. Still, there are signs that the ACLU is, to put it mildly, acting in a way that's uncharacteristic for a civil liberties group. As Kaminer has recently written, the ACLU had also tried to establish other oppressive internal regulations.

"Like the proposal governing board members' rights to speak, the agreements nearly imposed on the staff (but withdrawn after they became public) included a virtual gag rule; they also would have required the staff to acknowledge that all their communications on ACLU systems were subject to surveillance." [15]

Surveillance of staff computers and e-mails by the ACLU? Are we missing something here? Compiling files on internal political enemies, attempting to stifle free speech that is critical of the organization or its staff, data mining for prospective donors, purging dissidents who dare to disagree . . . Is this what the ACLU has become? It sounds more like the Communist Party.

Ironically, spying and surveillance—by others—is another serious concern of the ACLU. The ACLU's website contains a short video entitled *The Spies Have It*, urging citizens to protest the unauthorized tapping of phones and reading of e-mails. It also urges contacting corporations to seek a "No-Spy Pledge."

Perhaps the ACLU should consider taking the "No-Spy Pledge" itself.

Although the policies Meyers and Kaminer objected to have been resoundingly defeated because of the widespread negative publicity, Romero has ultimately prevailed. Meyers was unseated from the board, and Kaminer eventually decided against running for reelection. With Romero's sharpest critics effectively silenced, it remains to be seen if anyone will challenge him if his policies should once again run afoul of the ACLU's core principles.

Romero is not apologetic about his battles with board members. When asked by Josh Gersten of the *New York Sun* about an earlier statement by Mr. Meyers that Romero had told him and other critics to "kiss my Hispanic ass,"[16] he corrected him.

"I think it was, 'Kiss my Puerto Rican ass,'" Mr. Romero said matter-of-factly."[17]

The ACLU doesn't have a clue. And that's an outrage.

THE 2006 CONGRESS:
No Wonder We Threw Them Out . . . and How to Make Sure They Don't Go Back to Their Old Tricks Again!

On Election Day 2006, the Republican congressmen and senators of the 109th Congress watched in stunned disbelief as they lost dozens of congressional seats they had come to view as permanently theirs. Suddenly, everything had changed. For the first time since 1994, both Houses were now controlled by the Democrats, with Nancy Pelosi as the new speaker and her counterpart, Harry Reid, as the Senate's new majority leader.

The Republicans had good reason for their bewilderment. Over the years, with the full complicity of the Democrats, they had carefully gerrymandered their congressional districts to make it almost impossible to defeat incumbents. As a result, they complacently assumed they had achieved that rarity usually reserved for only monarchs and Supreme Court justices: lifetime tenure in political office. But by November 2006, the voters had caught on and were in no mood for the arrogance their leaders had come to display.

For most of the previous decade, the Republican majority in Congress had grown confident that they had hoodwinked the voters and immunized themselves from constituent wrath by a combination of their massive fund-raising operations, constant subsidized mailings, huge staffs, and safely drawn districts. Things had grown so stagnant in Congress that members in both parties—including many who survived the Massacre of '06 and are still in office today—were content to plunder everything that wasn't nailed down.

They became blatantly beholden to special interests—using their corporate jets to take them to both personal and political destinations, traveling all over the world at the expense of organizations that lobby Congress, and soliciting and accepting enormous campaign contributions from corporations seeking special treatment in federal legislation.

At the same time, in order to accommodate the time demands for fundraising and ingratiating lobbyists, they dramatically decreased the number of days they actually worked, directing more and more of their time to matters and events of importance to the top lobbyists without any relationship to the needs of their constituents.

It didn't stop there. They hired their spouses and other family members to work on their campaigns. In some cases, the family members were actually paid a percentage of the campaign donations made by lobbyists and other donors. And many of the heavy-hitting D.C. lobbying firms employed the spouses and family members of influential members to lobby their own husbands, wives, brothers, sisters, and nieces. It was a free-for-all.

It's not surprising then that voters began to believe that members of Congress had forgotten what they were elected to do. On Election Day, the voters reminded them: thirty-one members of the House and six members of the Senate were defeated.

See the sidebar for a short list of the abuses that may have sent so many members back home.

ABUSED BY THOSE WE ELECT:

HOW CONGRESSMEN AND SENATORS RIP US OFF

- The 109th Congress stayed in session for only 103 days in 2006 and still expected to collect full-time pay and benefits—despite going months and months without doing any official work.

- Instead of using their time to address legislation, our senators and congressmen took hundreds of free, all-expense-paid trips—paid for by lobbyists and other groups—all over the world, to warm places in the winter and Europe in the summer. Most took their wives; some have even taken their siblings or children for free!

- Our lawmakers took campaign contributions from lobbyists, and in return made sure that their clients got special secret earmarks of federal money—our tax money—for projects that, in many cases, constituents neither want nor need.

- They flew around the country on private corporate jets, accompanied by lobbyists seeking political favors, and only paid the cost of an equivalent first-class ticket, a fraction of the real cost of such flights. (Of course, even those payments are covered by their campaign committees, not the candidates themselves.)

- Dozens of them figured out a way to scam the system by turning campaign contributions into personal income by putting their husbands and wives on their campaign payrolls.

- More and more spouses, sons, daughters, sons-in-law, daughters-in-law, brothers, and extended family members are registering as lobbyists in Washington, DC. These new lobbyists can make a mint lobbying Dad, Mom, their niece, or big brother. Take former House speaker Dennis Hastert: his son closed his music store, moved to Washington, put out his shingle as a lobbyist, and somehow landed Google as a client. The new Senate majority leader, Harry Reid, has several sons and a son-in-law who have worked as lobbyists!

- Congress has become its own revolving door, where former members become lobbyists, trading on their expertise and contacts while they await their next turn in office.

It's time to change all this. We need to insist that the new leaders of Congress pass *permanent* reforms that will stop either party from ripping us off—now and in the future.

Sure, there's been a sweep of Congress. But can the new members manage this kind of full housecleaning? That remains to be seen. With the Democrats controlling the leadership in both houses, will they meet the challenge or simply take possession of the cookie jar and help themselves?

Have they learned the bitter lessons of the past six years and proved that they're prepared to halt the commercialization of Congress? Will they rein in the abuses and restore the legislative branch to serve the people and promote the ideal of public service? Let's hope so.

Both Houses have begun making significant changes, with bipartisan support. But there is much more to be done.

It should be—and it is—a privilege and an honor to be a member of Congress. But it's also a pretty good deal. Very few jobs in the private sector offer the perks, freedom, and opportunities given to our elected representatives. If congressional representatives and senators had real jobs, the help-wanted ad for their positions would read like this:

POSITIONS AVAILABLE

Members of House of Representatives and Senate

- For Senate: must be 30 years or older
- For House: must be 25 years or older
- Salary: $162,500 (plus annual cost-of-living increase)
- Extraordinary benefits: short work week, generally 2–3 days; health care; pension; 4–6 months paid vacation; opportunity for free travel with family; possible additional related employment opportunities for spouses and/or children
- Must run for office in federal election system

Sound good? That's the deal that we provide for members of Congress. They don't do much work, but they still get great perks. And we're the ones paying for it.

In 1948, President Harry Truman taunted the legislative branch by aggressively campaigning against the "Do-Nothing Congress." Unfortunately, not much has changed since then. Almost sixty years later, the U.S. Congress is quite similar to what it was in Harry Truman's time.

THE NEW DO-NOTHING CONGRESS

By any definition, Congress does not work hard. In fact, the 109th Congress dramatically decreased its agenda and workload by deliberately cutting back on the number of days that floor votes are scheduled in Washington. In 1948, the "Do-Nothing" House of Representatives denounced by Truman was in session for 108 days.[1]

In 2006, that same body spent only 103 days in session! In fact, from January 1 through June 30, the House held votes on only fifty-four days, while the Senate was in session for fifty-five days. Congress adjourned on October 4, leaving plenty of time for incumbents to campaign for reelection while still on the federal payroll. Then they gave themselves almost two more months off to recover from their exhausting work schedule. After the election, they actually came back in session for a few days, but didn't even really try to accomplish anything.

At that rate, of working only one day out of every four, the 2006 Congress was fixing to make Truman's Do-Nothing Congress look like a chain gang. No wonder we dumped them!

There is some hope of reform. Those who were fortunate enough to return to Congress understand that their constituents and the media are watching. Not surprisingly, one of the first initiatives taken by new House Speaker Nancy Pelosi was to announce that she was going to end the scheduling abuses, insisting that the lower chamber stay in session five days each week. That's good news to anyone who saw how little work Congress did under the Republicans.

Of course, by "five days," she actually meant from 6:30 PM on Monday to about 2:00 PM on Friday. Not exactly a traditional five-day week, but still an improvement. Theoretically, that is. As of the end of January 2007, there hadn't been a single five-day workweek in the House, even counting on the Pelosi clock. Even so, some of Pelosi's Democratic members were not too happy about the proposed infringement on their free time. Representative Deborah Wasserman-Schultz (D-FL) told the *Washington Post* she was going to have to reschedule her daughter's Brownie troop meetings that she leads on Monday afternoons to a weekend time.[2]

Earth to Congresswoman Wasserman-Schultz: That's what working women do in the real world!

But no sooner had Pelosi made the announcement than the House reverted to its old ways and announced that it would only meet for four days during the week of January 8. Why? A national emergency? A catastrophic storm? No. The collegiate championship football game between Florida State and Oklahoma State. House Majority Leader Steny Hoyer explained that he wanted to afford the members of Congress from the competing states the opportunity to watch the game. The equivalent of a congressional snow day! When it was the *Republicans* who were scheduling eight-day months, Hoyer was outraged. Once the Democrats controlled the calendar, he changed his tune. Suddenly a college football game started looking like a pretty important event:

> *There is a very important event happening Monday night, particularly for those who live in Ohio and Florida. In the spirit of comity, and I know if Maryland were playing, I would want to be accommodated and I want to accommodate my friend, Mr. Boehner.*[3]
>
> —Steny Hoyer, referring to new House Minority
> Leader John Boehner (R-OH)

The absence of any conventional work schedule in Congress for most of the past decade is truly shocking and needs to change. The early days of the 110th Congress are an improvement, but they still don't in any way resemble the work schedule of most Americans. Even with the Pelosi reforms, the House was scheduled to work only sixteen days in January 2007, the Senate seventeen. As the session evolved, the work week went back to the two to three day schedule.

And the quality of their work, in some cases, leaves a lot to be desired. An example: according to the clerk's record of roll call votes on Monday, January 29, 2007, the House considered its first vote at 6:56 PM (after returning from a four-day weekend). The vote was on a bill to designate a Post Office in Rock Island, Illinois, as the Lane Evans Post Office Building. The next bill, to designate a Vail Colorado Post Office as the Gerald R. Ford Post Office Building, was passed at 7:07 PM. The final bill of the day, considered at 7:16 PM, commended the University of Louisville Cardinals for winning the Orange Bowl (football again—apparently a big thing in the

House). Twenty-seven members were absent from the very short—twenty-minute—proceedings.

What is it with these members of Congress? While the rest of us are working to pay their salaries, they're spending their time doing everything but what they're elected to do. Instead of working on legislation in the myriad of public policy areas that cry out for their attention, they've been taking free vacations, raising money for their own reelections, campaigning on company time to stay in office and keep their generous perks, while generally skimping on their constitutional duties.

They weren't elected to be fund-raisers, but that's how they spend a lot of their time. They weren't elected to spend their work time campaigning to keep their jobs, but that's what they do. They weren't elected to go on expensive foreign junkets, but just watch them go.

They must not even like being in Washington too much because they're hardly ever there anymore. In all of 2006, there wasn't a single month when Congress went without at least a week's vacation. And, aside from those frequent respites, they were out of session altogether in January, August, October, and November and a few days in October and December 2006. And 2007 looks like it will only be a slight improvement.

Of course, the congressmen/women and senators will claim that their time away from Washington isn't really vacation time at all, that they're actually working hard in their districts. Sure. The problem is that the work they're doing generally has to do with making sure they get reelected, not tending to their constituents' needs.

Take a look at the outrageous 2006 congressional schedule.

EASY STREET: CONGRESS-STYLE 2006

JANUARY	In session only one day for one half hour: January 31
FEBRUARY	In session two days per week for three weeks. The rest of the time? The "Presidents' Day District Working Period"

(continued)

MARCH	Nine working days. From 3/16 through 3/28, though, time for a break: "St. Patrick Day District Working Period"
APRIL	Six days of work. Then a well-earned "spring vacation," April 6 through 25
MAY	Heavy work: thirteen days in session. But then a very long Memorial Day recess, from May 25 through June 10
JUNE	Sixteen days in session
JULY	Twelve days in session, with a break for Independence Day, from June 29 through July 10
AUGUST	Not in session *at all*. Summer recess.
SEPTEMBER	Return on 9/6; fourteen days in session
OCTOBER	Four days in session; then campaign season begins
NOVEMBER	Two days in session
DECEMBER	Five days in session

When we say "no sessions," we mean it: no votes, no hearings, no committee meetings. No nothing.

In general, votes are also carefully scheduled so as not to unduly inconvenience what is euphemistically called the "Tuesday to Thursday" club. In the House, at least, many of the votes are slated to begin only after 6:30 PM on Tuesdays and to end by Thursday. That way there's not too much interference with weekend plans.

REFORM? THE 2007 HOUSE SCHEDULE CALLS FOR TWENTY WEEKS AWAY FROM CONGRESS!

The new Democratic majority in Congress claims to be reforming the House and Senate scheduling abuses. But a closer look at their official schedules for 2007 shows otherwise. As of late January, the House had already blocked off twenty weeks, and the Senate nineteen, for either "District Work Periods" or adjournment, including most of August. That's 100 days. Add weekends, when they never work, and you start with a total universe of 161 possible session days. With additional holidays, we're down to about 155. Add weekdays when they are not actually in session, and the workweek is likely to average three days a week or less in 2007. In January, for example, the House was in session for a total of sixteen days, five days less than the rest of the full-time workforce—even counting New Year's Day and Martin Luther King Day.

Some would argue that the "District Work Periods" are a euphemism for either vacation time or personal fund-raising time, especially because it usually coincides with holidays like Presidents' Day, Easter, Memorial Day, Fourth of July, and just about all of August. But one thing is for sure—there's no work at all done in Washington during that time.

HERE'S THE HOUSE NON-SESSION DAY SCHEDULE FOR 2007:[4]

- January 1: New Year's Day
- January 15: Martin Luther King Day
- February 19–26: "Presidents' Day District Work Period": one workweek off
- April 2–13: "Spring District Work Period": two workweeks off
- May 25–June 4: "Memorial Day District Work Period": one workweek off
- June 29–July 9: "Independence Day District Work Period": one workweek off
- August 6–September 4: "Summer District Work Period": four workweeks off
- September 13: Rosh Hashanah
- September 22: Yom Kippur
- October 8: Columbus Day
- October 26–January 2007 adjourned: eleven workweeks off!

The Senate schedule for 2007 isn't much better; they'll be in session for only a few additional days. In the month of January 2007, they worked a total of seventeen days, while the rest of the working world put in twenty-one or twenty-two. The official Senate schedule for January–September 2007 calls for one week off in April, instead of the two weeks planned by the House.[5] So, the Senate will be out of session for nineteen of fifty-two weeks this year! That's no reform.

Maybe the members should think about the thirty-plus former congressmen and six former senators who can really relax and spend full-time at home—because they were kicked out by the outraged voters! We need full-time legislators, not part-time travelers and fund-raisers.

Here's an easy reform that would let us know exactly what our leaders are doing every day: all members of Congress should post their schedules on their websites every day. That way, we voters could decide whether they're really working for us—or for themselves.

WHAT THEY DIDN'T DO

Everybody says the Republicans lost Congress because of the war in Iraq. But that's like saying that a person died of pneumonia when what really killed him was the AIDS virus that made him vulnerable to infection.

Yes, the Republicans lost over Iraq, but they really lost because the 109th Congress didn't do anything. With no record to defend, the incumbents found themselves on the defensive over Iraq and lost their seats.

The 109th Congress, which served from 2005 through 2006, is most notable for having achieved just about nothing. When it convened, in the heady days following Bush's reelection, it had two main items on its plate: Social Security reform and immigration.

President Bush proposed a major overhaul of Social Security, seeking, as he put it, to spend his "political capital" on the issue. His proposal called for allowing workers to channel some of their Social Security tax payments to individual accounts, where they could choose how to invest their money. But the plan went nowhere. The elderly were seized by dread that their pension money would be drained, even though Bush had specifically guaranteed that their payments would not be affected.

Recoiling in horror at the citizen outcry, Congress did the brave thing: they buried the bill in committee and never let it see the light of day on the floor, where they would have been compelled to vote on it.

Immigration reform didn't fare much better. The House adopted a tough bill calling for a border fence and an aggressive program to stop employers from hiring illegal immigrants. The Senate embraced many of the House's ideas but added an earned path to citizenship for those now here illegally. Again the Congress was frightened by the public outcry. Hispanic Americans demonstrated against the draconian House bill, and conservative groups railed against what they called "amnesty" in the Senate version.

So what happened? Once again, the leaders of Congress shied away from a fight and buried the bills in committee, never seeking to reconcile the different House and Senate versions. Finally, in the last days of the session, just before the election, the Congress passed legislation mandating a seven-hundred-mile fence on the border with Mexico. A good start, but the legislation didn't address a small problem: the 11.5 million illegal immigrants who are already on *our* side of the fence!

Two other issues cropped up during the session: energy and ethics.

Surging gasoline prices prompted calls for a tough energy bill, but Congress punted, enacting a tame piece of legislation that served mainly to give more benefits to oil and gas producers but did little to turn the country to alternative forms of energy or even to expand our production of oil, since drilling in Alaska's wildlife areas was rejected.

On ethics, the 109th did virtually nothing. Despite high-profile scandals involving lobbyist Jack Abramoff and Congressman Tom DeLay that exposed the sinister and close relationships between lobbyists and legislators, Congress neither tightened the laws nor enacted more disclosure requirements. At first, the focus on congressional travel seemed to be leading to a well-needed ban on privately paid travel. But that faded away after the initial unfavorable publicity died down. And they minimized their travel schedules in the last months of the session. Once the spotlight was centered on the extensive traveling by members, the number of trips drastically decreased.

It's not literally true that the 109th Congress did nothing at all. Proving Mark Twain's maxim that "No man's life, liberty or property is safe while the legislature is in session,"[6] Congress did pass two particularly reprehensible bills, raising interest rates on student loans and making bankruptcy a permanent condition, punishing debtors without relief or respite. (There's more on these two horrible laws to come—read on.)

To put it bluntly, the 109th Congress seemed to pour most of its energy into one area: feathering their own nests. From free trips to lobbyist campaign contributions to earmarked federal funds to hiring their wives on the payroll to lobbying by family members, it made new strides in lax ethics and legal corruption.

To get an idea of how utterly insignificant the accomplishments of the 109th Congress were, aside from mandatory appropriations, budget, and confirmation votes, consider this list of some of the House floor actions:

SOME THINGS THEY DID DO: ACCOMPLISHMENTS OF THE 109TH CONGRESS[7]

- Celebrating the 40th anniversary of the Texas Westerns 1966 NCAA Basketball Championship and recognizing the groundbreaking impact of the title game victory on diversity in sports and civil rights in America

- Establishing the Thomas Edison National Historical Park in the State of New Jersey as the successor to the Edison National Historical Site

- Honoring the contributions of Catholic schools

- Recognizing Hall of Famer Bob Feller

- Expressing the sense of the House of Representatives that a National Young Sports Week should be established

- Recognizing and honoring the 100th anniversary of the founding of the Alpha Phi Alpha Fraternity, Incorporated, the first intercollegiate Greek-letter fraternity established for African Americans

- Recognizing the importance and positive contributions of chemistry to our everyday lives, and supporting the goals and ideals of National Chemistry Week

- Expressing the sense of Congress regarding the education curriculum in the Kingdom of Saudi Arabia

- Expressing the sense of the House of Representatives regarding the study of languages and supporting the designation of a Year of Languages

- In eleven separate bills, changing the names of post offices, federal buildings, and courthouses to honor certain persons

- Directing the secretary of the Interior to conduct a pilot program under which up to fifteen states may issue electronic federal migratory bird hunting stamps

- Congratulating the National Football League champion Pittsburgh Steelers for winning Super Bowl XL and completing one of the greatest postseason runs in professional sports history

- Celebrating the 50th Anniversary of the International Geophysical Year and supporting an International Geophysical Year 2 in 2007–08

Of course, we have no desire to take anything away from Thomas Edison, Bob Feller, or the International Geophysical Year. We just thought it might not be unreasonable to expect a little more from our hard-working elected representatives.

FLEEING THE COOP

If they can't really be called full-time legislators, a lot of our congressmen and senators can definitely be described as nearly full-time *travelers*.

The enormous amount of free time they spend away from Washington provides members of Congress the opportunity to spend extensive travel time, to take their spouses on luxurious free trips to European capitals, Caribbean islands, the Middle East, the Far East, even to America's own premier resort destinations. At best these trips are financed by foundations, at worst by special interest groups. In the two years of the 109th Congress, almost five million dollars was spent on congressional travel by private organizations.[8] This has been fairly consistent for at least the last six years—as the website www.politicalmoneyline.com chronicles in troubling detail.

Luxurious Travel: The Ultimate Congressional Perk

Since January 2000, more than $20 million has been spent by private organizations on luxurious travel for members of Congress and their spouses.[9]

You read that right: TWENTY MILLION DOLLARS!

Travel has become a key perk of a job in Congress. There are organizations of all kinds lining up to shower the members with free trips. Some do so to curry favor, to help ensure that they can ask for—and receive—legislative favors in return. Other interest groups do it so they can have their lobbyists sit next to the congressman and personally fill his ears with their side of the story. The goal is to build a personal relationship that can be called on in the future.

Some trip sponsors have no specific legislative agenda, but offer "educational" trips, raising foundation money for the purpose. Whatever the motive, this freebie travel has transformed the Congress into a kind of see-the-world club, where members get to live imperial lives in exotic places. The travel is so pervasive that it's become an institutional perk, one of the core benefits of being elected to Congress.

To its credit, upon entering office in early 2007, the 110th Congress immediately enacted serious reforms that regulate and restrict member travel. Both the House and Senate have now banned all travel paid for by lobbyists or the firms that employ them.[10] In addition, the Senate has passed a blanket prohibition of the use of corporate jets by senators. However, the House didn't go that far. The new House rules now require members using corporate jets to reimburse the company for the actual price of the trip, not just the full first-class airfare. That's a big difference in cost, but it's still no big deal for the members. They will simply charge the higher cost to their campaign committees and continue to fly in the lap of luxury, while corporate lobbyists host them as a captive audience. This is not a reform—it's a phony accommodation. Instead, the House needs to follow the Senate's lead and ban the use of corporate jets by lawmakers under any circumstances. There is no public good that can come from this practice.

The truth is, neither chamber has gone far enough on regulating travel.

All Privately Funded Travel
by Members of Congress
Should Be Prohibited

Congressmen are elected to work on legislation, not to receive gifts of luxurious travel, regardless of the source. Besides the contributions of lobbyists, members of Congress spend too many days on exotic vacations paid for by other groups—another kind of perk that should be banned outright. These trips take lawmakers away from Washington and away from their districts for no public purpose. Any meaningful seminars could be held just as easily in Washington, D.C.

There's one other reform that's desperately needed:

**Any travel by any member
of either house should be
documented and available electronically
for constituents to review
in an easy and organized format.**

Right now, for example, it's impossible to see records of government-funded travel by senators without physically going to the Capitol and reviewing and copying files by hand. However, records of privately paid trips are now being made available by several organizations that provide an important and necessary service.

HOW TO GET THE TRUTH ABOUT YOUR CONGRESSMAN, TO CONNECT THE DOTS BETWEEN CAMPAIGN CONTRIBUTIONS AND LOBBYISTS

Political Moneyline: http://www.PoliticalMoneyLine.com

Public Citizen: http://www.citizen.org

Center for Responsive Politics: http://www.opensecrets.org

(continued)

Sunlight Foundation: http://www.sunlightfoundation.com

Medill News and Center for Public Integrity: http://www.medillnewsdc.com/power_trips/trips06_aboutpowertrips.shtml

American Radio Works: http://americanradioworks.publicradio.org/features/congtravel/

These organizations provide invaluable information and insight into everything Congress wants to keep a secret. Check with them regularly to stay current about what your member of Congress is up to. (You won't find this information on their official websites!)

THE BIGGEST TRAVELERS IN CONGRESS, 2000–2006

	Cost
James Sensenbrenner (R-WI)	$210,786
James A. McDermott (D-WA)	197,572
George Miller (D-CA)	192,166
Raymond E. "Gene" Greene (D-TX)	184,329
Phillip S. English (R-PA)	182,297

Source: Politicalmoneyline.com: http://www.politicalmoneyline.com/cgiin/x_Private Detail.exe?DoFn=BiggestSpendersPoliticalmoneyline

ONE MONTH'S TRAVEL: PORTRAIT OF A DO-NOTHING CONGRESS

January isn't normally a month Americans associate with vacations. It's usually a time when we get back to work after the Christmas and New Year's holidays. But to understand just how much traveling members of Congress did last year, it's interesting—and horrifying—to look at what they did and where they went in January 2006, while the rest of us were busy working.

In all, members took 160 privately paid trips during that single month—almost always accompanied by their spouses, and sometimes by a sister, brother, or child. These trips had only one thing in common: the member of Congress didn't pay anything at all for them. Some of the highlights are noted in the sidebar.

Seven House members—five of them women—spent a week or more in **Israel** as guests of the American Israel Education Foundation. Their costs included $1,500 per person for security, entrance fees, and, in most cases, "honoraria." None of them were members of the House Committee on International Relations. (The data below were compiled by www.politicalmoneyline.com and are based on the official travel reports filed by the members and their travel sponsors.)

- **Tammy Baldwin** (D-WI) and her partner, Lauren Azar, traveled to Israel from January 8–15. The couple added several additional vacation days to their trip in order to tour Jordan at their own expense from January 4–8.[11] Total cost of the Israel portion: $17,908. House Rules permit members to bring family members; Ms. Baldwin identified her partner as her "spouse" on her disclosure form

- **Diana Degette** (D-CO) traveled to Israel from January 7–15. Cost: $8,647

- **Judy Biggert** (R-IL) and her husband also went to Israel at a cost of $15,722

- **Virginia Brown-Waite** (R-FL) brought her sister, Barbara Stearns, to Israel with her from January 7–15. Total cost: $18,381

- **Sheila Jackson-Lee** (D-TX) traveled to Israel and Gaza from January 8–14. Total cost: $17,237. Although she did not disclose a spouse or other guest, the amount of the airfare ($9,927) as well as the costs for security, etc., suggest that another party joined Ms. Jackson-Lee

- **Henry Waxman** (D-CA) also traveled to Israel from January 4–14. He and his wife were the guests of the American Jewish Congress and they added five additional days to their trip for a vacation. Total cost: $9,290

- **Tim Ryan** (D-OH) spent January 6–13 in Israel as the guest of the Jewish Community Board of Akron, Ohio. Total cost: $5,755

Three House members traveled to China as guests of the National Committee on U.S.-China Relations from January 10–18. None were members of the House International Relations Committee. They were **Tom Feeney**

(R-FL), for a cost of $10,804; **Mark Kirk** (R-IL), for a cost of $10,665; and **Rick Larsen** (D-WA), for a cost of $10,732.

Three Congressmen, none of them members of the House International Relations Committee, traveled to Eastern Europe as guests of the "International Management and Development Institute." They were **Robert Aberholt** (R-AL), and his wife, who spent a week traveling in Vienna, Budapest, and Kazakhstan for a cost of $18,434; **James McDermott** (D-WA), and his wife, who also traveled to Kazakhstan for a "government business dialogue with Kazakhstan government officials and business leaders" [12] from January 13–18 for a cost of $11,660; [13] and **Maurice Hinchey** (D-NY), who traveled to Vienna, Budapest, and Kazakhstan from January 10–18 for a cost of $9,794.

Four House members were the guests of the Inter-American Economic Council in Montego Bay, Jamaica, for participation in a "Business Roundtable": **G. K. Butterfield** (D-VA) spent five days in Jamaica for a cost of $4,771.00; **Albert Wynn** (D-MD) and his wife traveled to Jamaica from January 11–15 for a cost of $9,547; **Gregory Meeks** (D-NY) and his wife spent five days in Montego Bay from January 11–15 for a cost of $9,747; **Stephanie Tubbs** (D-OH) also attended the Jamaica conference, for a cost of $6,129. (From January 17–21, Ms. Tubbs also visited Barbados, West Indies, for the Judicial Conference of the National Bar Association, for a total cost of $1,340.)

Seven members of Congress went to Kona, Hawaii, for a week in January as guests of the American Association of Airline Executives to participate in a conference on "Aviation Issues":

Rep. Harold Dallas Rogers (R-KY) and his wife, who added two extra days at their own expense, for a cost of $7,209

Sen. Trent Lott (R-MS), for a cost of $6,517

Rep. Daniel Lungren (R-CA), and his wife, for a cost of $8,546

Rep. Bennie G. Thompson (D-MS), for a cost of $9,545

Rep. Martin Sabo (D-MN), for a cost of $11,400

Rep. Edward L. Pastor (D-AZ), and his wife, for a cost of $2,002

Rep. Joseph Knollenberg (R-MI) and his wife, for a cost of $7,041

Five members were guests of the Association of American Railroads in Fort Myers, Florida:

- **Rep. James Clyburn** (D-SC) $4,761

- **Rep. Corinne Brown** (D-FL) $4,168

- **Rep. Eddie Bernice Johnson** (D-TX) $3,780

- **Rep. John Duncan** (R-TN) $3,780

- **Rep. Bennie Thompson** (D-MS) $1,334

Fifteen members and their spouses were the guests of the Aspen Institute in Punta Mita, Mexico, to discuss U.S. policy in Latin America:

- **Rep. Ed Markey** (D-MA) and wife $7,805

- **Rep. Edward Berman** (D-CA) and wife $5,852

- **Sen. Barbara Boxer** (D-CA) and husband $6,739

- **Rep. Susan Davis** (D-CA) and husband $7,305

- **Rep. Norm Dicks** (D-WA) and wife $7,689

- **Rep. Sam Farr** (D-CA) and wife $6,695

- **Rep. Raymond "Gene" Greene** (D-TX) and wife $6,164

- **Sen. Tom Harkin** (D-IA) and wife $7,117

- **Rep. Sander Levin** (D-MI) and wife $6,123

- **Rep. Nita Lowey** (D-NY) and husband $7,327

- **Sen. Richard Lugar** (R-IN) and wife $7,470

- **Rep. George Miller** (D-CA) and wife $6,950

- **Rep. Janice Schakowsky** (D-IL) and husband $7,045

- **Rep. Tom Udall** (D-NM) and wife $7,567

- **Rep. Roger Wicker** (R-MS) and wife $9,067

Three members of Congress traveled to Charleston, South Carolina, for the "Renaissance Weekend":

- **Rep. Doris Matsui** (D-CA) $3,166

- **Rep. Chris Shays** (R-CT) $2,450

- **Sen. Hillary Clinton** (D-NY) $380

And the following members also traveled:

- **Rep. Chris Shays** (R-CT) went to Africa as a guest of Planned Parenthood at a cost of $9,509

- **Rep. Melvin Watt** (D-NC) traveled to Barbados for the National Bar Association at a cost of more than $2,000

- **Rep. John Linder** (R-MD) and his wife were guests of the Ripon Society (a centrist Republican think tank) in Palm Beach, Florida, at a cost of $4,084

- **Rep. Pat Roberts** (R-KS) and his wife were also guests of the Ripon Society in Palm Beach at a cost of $4,510

Amazing, isn't it?

And there were other trips in January, but not of the significance of these. And, remember, there was one half-hour session of Congress during that time. Do you see why it was time for reform?

FREE TRIPS: WHAT FUN IT USED TO BE!

For a good example of how our elected officials traveled before the recent reforms, take a look at how Congressman Harold Rogers of Kentucky and his wife benefited from the airline, railroad, and homeland security companies who have business before the committees he serves on.

Congressman Hal Rogers: Working for the Railroads, All the Livelong Day

Harold Dallas Rogers (R-KY), who survived the 2006 massacre, is a member of the Appropriations Committee and its Subcommittee on Transportation, Treasury, and Independent Agencies.

The key word here is *transportation*. The transportation industry is very generous to Rogers, who oversees legislation relating to railroads, airlines, and highways. In the 2006 election cycle, Rogers received $76,000 in PAC contributions from the transportation industry and more than $49,500 from defense industries.[14] Rogers is also chairman of the Appropriations Subcommittee on Homeland Security.

Although he was virtually unopposed in 2006, he raised $1,042,711. His Democratic opponent raised a total of $1,624. That's right—less than two thousand dollars. Did anyone really consider this a serious race?

The transportation special interests didn't stop at campaign contributions alone. They wanted Rogers to enjoy life, too. Since 2000, Rogers and his wife Cynthia have spent over 125 days on twenty-nine separate trips.[15] Twenty-six of those, costing over $150,000 in total, were financed by the airline, railroad, transportation, and homeland security industries.[16] Based on data compiled by www.politicalmoneyline.com, Rogers traveled on the dime of the following organizations:

- **The Association of American Railroads** sent Hal and Cynthia to Palm Springs, Wyoming, Orlando, Washington, Fort Lauderdale, and San Diego

(continued)

- **The American Association of Airline Executives** brought the couple to Hawaii on six different trips, totaling over forty days, as well as other trips to San Diego, San Francisco, Palm Springs, and Dublin. The Rogerses traveled for a total of more than sixty days—two months—as the guest of the airline executives

- **Burlington Northern Santa Fe Railroad** (BNSF) hosted the couple in Wyoming

- **Greater Orlando Aviation Executives** brought them to Orlando

- **Dallas Area Rapid Transportation** brought them to Dallas

- **CSX Corporation,** an international transportation company, hosted them at the Greenbriar Hotel in West Virginia

- **PRSM Corporation,** a company dealing in homeland security products, invited them to Tennessee; another such company, **Securimetrics,** brought them to Sun Valley, Idaho

Between 2000 and 2006, Congressman Rogers also visited the following countries—at a *government* expense of well over $30,000: France, Italy, Turkey, Australia, New Zealand, Croatia, Spain, Greece, Ireland, Norway, and Columbia. (Source for all of the above: www.politicalmoneyline.com.)

And then There Was Ireland . . .

Of special interest was a trip Rogers took in August 2003 for five days in Ireland, followed by five days in London. In the initial filing of the required form disclosing the trip, Rogers and his fellow travelers all indicated that the trip was paid for by Kessler and Associates, a lobbying firm that is prohibited from paying for congressional trips. No one focused on it until two years later, when a reporter for *The Hill* questioned the legality of the expenditures by a lobbying organization. Suddenly, all of the travelers played musical chairs and changed the form to attribute the travel to Century Business Services, Inc., the parent company of the Kessler firm. That trip was the only foreign trip that Century has ever paid for.

While it's hard to tell which one of his special interest clients Kessler was especially promoting on the trip, he definitely "represented drug companies Pfizer, Amgen, Bristol-Myers Squibb, and Abbott Laboratories, in addition to cigarette manufacturer Altria, electric utility company Exelon, and BNSF Railroad."[17]

Is there any doubt that the special interests really paid for the trip?

Accompanying Rogers on the excursion were **Sen. Gordon Smith** (R-OR), former **Sen. Don Nickles** (R-OK), and **Reps. Howard Coble** (R-NC), and **Clay Shaw** (R-FL). The trip cost approximately $10,000 per couple.[18] Congressman Shaw was defeated in the 2006 election. Now he'll have to pay for his own travel.

Incidentally, all of those folks were also well taken care of: Shaw received $120,600 in contributions from the transportation industry, Coble received $48,000, and Smith was given $3,500.[19] (Nickles did not run again in 2004.)

NOTE: The new Democratic Chairman of the Committee on Transportation and Infrastructure and Railroads, James Oberstar, received $165,656 from the transportation industry.

Back to Ireland: This trip was an especially luxurious one. Lobbyist Richard Kessler accompanied the five members of Congress and their wives who stayed at Ashford Castle, an elegant and expensive thirteenth-century castle turned five-star hotel perched on Ireland's second largest lake. The former home of the Guinness family, the castle offers golf, tennis, fishing, horseback riding, and a state-of-the-art spa and beauty salon. In 2006, it was voted as one of the top hotels in Europe by *Travel and Leisure* magazine. That kind of luxury doesn't come cheap: the lowest rates begin at about $500 per person per night. The hotel's website describes the luxurious setting:

> Ashford Castle has an unsurpassed reputation for providing guests with the very finest rooms and suites. Each guest room is individually designed to provide stylish personal comforts, from the marble fittings of bathrooms to sumptuous co-ordinated fabrics and furnishings.[20]

The premises include four restaurants and two bars with the main dining room, the George V, named after the former Duke of Wales—who visited the house before he became King George V. Its standards are certainly befitting a duke, a king, and, apparently, even a congressman.

Those beautiful August days weren't an especially busy time for the travelers; there was plenty of time to relax and enjoy the posh surroundings. According to *The Hill,* "The group spent thirteen hours from August 4 to 8 in roundtable discussions on subjects ranging from 'TEA 3-U.S. Transport Policy Approaches for the 21st Century' to 'Intellectual Property and the Protection of Enterprise,' according to a draft schedule of activities printed on Kessler & Associates letterhead." [21] Over a five-day period, that's about two and a half hours a day.

After leaving Ashford Castle on August 9, all five lawmakers attended the Ripon Society's 21st Annual TransAtlantic Conference in London, noting the event in their records as a separate occasion sponsored by the "Ripon Educational Fund," [22] a not-for-profit arm of the Ripon Society. The founder of the Educational Fund is, coincidentally, Richard Kessler. Ten of its twelve board members are lobbyists.

In London, fourteen other legislators joined the conclave. The purpose of the London shindig was to subject the new larger group of nineteen senators and congressmen to the persuasion, importuning, flattery, enticements, and attentions of more than one hundred lobbyists who outnumbered their targets by five to one, according to Public Citizen:

> While the [Ripon] Fund's sole purpose seems to be to organize and finance fancy conclaves between lobbyists and members of Congress, its non-profit educational status provides a convenient cover for Kessler and his fellow lobbyists. Lawmakers get to go on luxurious trips, accompanied by family members most of the time, while the lobbyists get to accompany them and pay the bill. [23]

The good news: the new 2007 reforms prohibit all such gifts of travel by lobbyists and corporations! No more free vacations paid for by special interests.

The bad news: there are plenty of other ethical violations out there that the reformers haven't caught yet.

ONE MORE PROBLEM: FOUNDATIONS AND OTHER GROUPS ARE STILL UNDERWRITING OUR LEADERS' EXPENSES

Although lobbyist funding of congressional travel is now prohibited, the abuses of the last decades are a shocking reminder of the need for constant oversight over congressional perks. And there's still another lingering problem: the widespread funding of expensive travel by non-lobbyist organizations. The need for reforms in this area has been virtually ignored. In fact, in early January 2007 the Senate passed an amendment as part of its ethics reform bill that specifically *permits* travel paid for by foundations and not-for-profit organizations so long as the travel is preapproved by the Senate Ethics Committee. This is not a reform—it's an outrage!

The problem is that millions of dollars are spent by these foundations and not-for-profits to allow our government officials to luxuriate in desirable vacation spots worldwide—wasting their time while helping to foster the sense that our "imperial" members of Congress deserve such treatment simply because they are elected officials.

If that upcoming seminar in a tropical paradise were really so important, Congressman, shouldn't they be holding it in Washington?

The Congressman's Best Friend: The Aspen Institute and the Gift That Keeps on Giving

By far the biggest spender on congressional trips is the Aspen Institute. Since 2000, they've paid almost $4 million for 699 trips, sending lawmakers and their spouses to glamorous places all over the globe.[24] That's deceptive, though, because the institute is not in the influence-peddling business—which makes them unique in Washington. Unlike most of the underwriters of congressional boondoggles, Aspen does not lobby Congress on any issue, does not accept funding from any lobbyists, and does not permit lobbyists to attend their conferences. Those are BIG differences from some of their scheming counterparts.

According to former Democratic senator Dick Clark, now Director of Aspen's Congressional Program, the conferences are designed to give members of Congress the opportunity to hear presentations by well-known scholars in important public policy areas. Clark emphasizes that the programs are politically balanced and that there is never just a single point of view presented. "I could never get Democrats and Republicans to come if I just offered them just one side of an issue."[25] In recent years, Aspen has sponsored programs on education reform, the politics of Islam, the politics of Latin America, Russian and U.S. relations, and China-U.S. relations.

Clark is right. Aspen is, indeed, in a totally different league in the purpose and focus of the travel it avails members of Congress. But there's still something troubling about the Aspen trips. While they ask nothing in return of the members of Congress who attend, they tend to invite the same people over and over. As a result, some members are constantly traveling to exotic places on Aspen's dime—a very valuable perk. Some of them extend their trips at their own expense, so that they have added vacation time in the foreign city. Once the airfare is paid for, the additional expense to the member is nominal. What would be a very expensive trip for anyone else becomes a highly affordable luxury to many members of Congress.

Doesn't it distort the perspective of those senators and congressmen who frequently accept Aspen's hospitality to have luxurious world travel so effortlessly at their disposal—simply because they've been elected to serve us in Congress? This perk surely contributes to the high-and-mighty lifestyle pretensions of our congressmen and senators today—pretensions our founders feared, and sought to avoid, in Congress. Should it be part of our representatives' jobs to take extended free vacations, required only to attend a few interesting panel sessions in return? Even if they aren't surrounded by lobbyists pushing private interests, what does this sort of free luxury do to our leaders' ideal of humble public service?

For some members of Congress, Aspen travel seems to have become a way of life. Congressman George Miller (D-CA), for example, has taken twenty-seven trips at the Institute's expense since 2000. In fact, Aspen does not seem to succeed in getting new members to come to its foreign conferences. It's usually the same old crowd. In the first seven months of 2006, for example, only one of the forty congressmen or senators who went on foreign trips was attending an Aspen program for the first time. In 2005, only five of the ninety-eight members who traveled at Aspen's expense were new participants. (For details on all of these trips, go to politicalmoneyline.com.)

According to Clark, there is no set protocol on how the invitations are made. He points out that there are many members Aspen would love to host, but who are not interested in attending.[26] Still, many of their old favorites were invited time and again, raising another question: what is the point of having the same people get together year after year to discuss the same issues?

The Aspen organization itself is doing nothing inappropriate. What *is* inappropriate is that so many members of Congress devote their abundant time and energy to these repeated foreign trips. Here's a little rogues' gallery:

The Top Aspen Travelers

1. GEORGE MILLER (D-CA) The king of Aspen Institute travel is Rep. George Miller (D-CA), who attended twenty-six vacation-conferences sponsored by Aspen from 2000–2006. He and his wife, Cynthia, traveled the world courtesy of Aspen, visiting

Naples, Florida	Barcelona, Spain (twice)
San Juan, Puerto Rico	Montego Bay, Jamaica (twice)
Vancouver, Canada	Rome, Italy
Prague, Czech Republic	Moscow, Russia
Grand Cayman	Cancun, Mexico
Florence, Italy	Venice, Italy
Helsinki, Finland	Dublin, Ireland
Punta Mita, Mexico (three times)	Istanbul, Turkey (twice)
Scottsdale, Arizona	Honolulu, Hawaii
China	Krakow, Poland

Miller also went to Naples, Florida, and Lanai, Hawaii, without his wife. **The total cost of these trips? More than $125,000.**

The Millers took full advantage of the wonderful opportunities afforded to them by Aspen. Once they reached their exotic destinations, they frequently stayed on to enjoy the sights. The Millers extended their trip to Prague and Barcelona for an additional three days apiece; stayed on in Vancouver for an extra two days; and remained in Istanbul for an extra two days. Mrs. Miller also extended the Venice trip for eleven days and went on to Paris.

All told, George Miller—who happens to be the third highest recipient of private travel in all of Congress—was traveling to or attending Aspen programs for 161 days over a six-year period. That's an average of twenty-six days a year!

When asked about the repeated invitations to Miller, Aspen's Congress project director, Dick Clark, praised Miller as "the brightest member of the House."[27] Clark indicated that he would continue to invite Miller in the future. Until this year, Miller was the ranking member of the Education and Workforce Committee. Three of the twenty-six conferences were on education issues. The others dealt with topics such as "The Future of Political Islam," "U.S.-Russian Relations," and other foreign-affairs issues.

Think about it: why would any organization pay for any member of Congress—even arguably its brightest member—to take twenty-six trips over a six-year period?

More important, does Miller have so much free time as a congressman that he can afford to travel so frequently and extensively?

Miller's government-paid travel record is equally impressive. From 2000 to 2006, he visited the following countries, spending another thirty-six days of work time on free travel:

Mexico	Lebanon
Cambodia (twice)	Israel (twice)
Vietnam (twice)	Jordan
South Africa	Iraq
France	Italy (twice)
Hong Kong	Sudan (twice)
Taiwan	Ghana
Laos	Liberia
Egypt	Cape Verde

While such government-funded travel presumably relates to congressional work, Congressman Miller does not belong to any committee dealing with foreign relations. And yet between 2000 and 2006 he spent almost two hundred days traveling, at a total cost of over $165,000 (excluding military transport in certain cases).

He sure likes to travel!

2. SENATOR RICHARD LUGAR (R-IN) Another major beneficiary of the Aspen largesse is Senator Richard Lugar (R-IN). He and his wife went on twenty-five trips sponsored by Aspen. They visited

Naples, Florida
Helsinki, Finland (twice)
Grand Cayman
Punta Mita, Mexico (five times)
Scottsdale, Arizona
London, England
Montego Bay, Jamaica (twice)
Rome, Italy
Moscow, Russia

Honolulu, Hawaii
Cancun, Mexico (twice)
Barcelona, Spain
Lausanne, Switzerland
Venice, Italy
Dublin, Ireland
Istanbul, Turkey
Krakow, Poland

Lugar also went to St. Petersburg, Florida without his wife.

Admittedly, Lugar was chairman of the Senate Foreign Relations Committee; no doubt he had much to offer, and perhaps even to learn, from scholarly panels. But what's he doing at education-reform programs? And, again, how does the chairman of the most prestigious committee in the Senate find the time to globetrot so often?

3. CONGRESSMAN HOWARD BERMAN (D-CA) is another perennial Aspen invitee. He and his wife attended eighteen Aspen travel programs, visiting

Grand Cayman (twice)
Helsinki, Finland (twice)
Vancouver, Canada
White Sulphur Springs, West Virginia
Puerto Vallarta, Mexico
Moscow, Russia
Barcelona, Spain

Venice, Italy
Shanghai and Beijing, China
Istanbul, Turkey (twice)
Punta Mita, Mexico
Honolulu, Hawaii
Aspen, Colorado
Krakow. Poland

Like the Millers, the Bermans stayed on to further enjoy the sights on several trips—for three extra days in Venice and an extra day in Aspen.

4. REP. DONALD PAYNE (D-NJ) attended fifteen Aspen programs and visited

San Juan, Puerto Rico
St. Petersburg, Florida

Helsinki, Finland (twice)
Florence, Italy

Punta Mita, Mexico	Brussels, Belgium
Scottsdale, Arizona	Nassau, Bahamas
Montego Bay, Jamaica	Aspen, Colorado
Cancun, Mexico	Honolulu, Hawaii (twice)

Payne's brother Bill accompanied him to Scottsdale, Montego Bay, Helsinki, San Juan, St. Petersburg, and most recently to Honolulu in April 2006. What possible rationale could explain a congressman inviting his *brother* to a foreign conference?

5. REP. HENRY WAXMAN (D-CA) and his wife were also favorites of Aspen, attending twelve programs and traveling to

San Juan, Puerto Rico	Barcelona, Spain
Florence, Italy	Lausanne, Switzerland
Prague, Czech Republic	Venice, Italy
Helsinki, Finland and Talinn, Estonia	Punta Mita, Mexico
Shanghai and Beijing, China	Montego Bay, Jamaica
Moscow, Russia	Istanbul

The Waxmans added four extra days in Istanbul, three extra days in Venice, four in Switzerland, four in Florence, and four in Prague.

6. REP. LLOYD DOGGETT (D-TX) and his wife traveled to eleven Aspen conferences, including

China	Prague
Istanbul (twice)	Vancouver
Spain	Mexico
Moscow	Helsinki
Zurich	Florence, Italy

The Doggetts frequently expanded their Aspen-paid trips, including adding more than nine days in Russia, two days in China, four in Istanbul, and four in Zurich.

7. REP. MAURICE HINCHEY (D-NY) traveled to eleven Aspen conferences, including

Mexico (three times)
Vancouver, Canada
Prague, Czech Republic
Helsinki, Finland
China

Rome, Italy
Grand Cayman
Montego Bay, Jamaica
St. Petersburg, Florida

8. CONGRESSMAN RAYMOND "GENE" GREENE (D-TX) traveled to ten Aspen events, including

Naples, Florida
St. Petersburg, Florida
Scottsdale, Arizona
Rome, Italy

Mexico (three times)
White Silver Springs, West Virginia
Switzerland
San Juan, Puerto Rico

9. SENATOR BARBARA BOXER (D-CA) and her husband took ten trips to Aspen conferences, visiting

San Juan, Puerto Rico
Grand Cayman
Florence, Italy

Dublin, Ireland
Punta Mita, Mexico (four times)
Barcelona, Spain

10. CONGRESSMAN NITA LOWEY (D-NY) and her husband attended ten conferences including trips to

Naples, Florida
Vancouver, Canada
Helsinki, Finland
Mexico (twice)
Hawaii

The Bahamas
Venice, Italy
China
Grand Cayman

She and her husband added vacation time to the trips: two extra nights in Vancouver, six in Venice, and seven in China at personal expense.

Enough to give you jet lag, isn't it?

Senator Clark of the Aspen organization conceded that, to the average person, it might look like certain members and their spouses are given repeated vacations in luxurious spots around the globe. But he insists that the sessions require serious preparation and participation of about six hours a day, including meals. Spouses are encouraged, but not required, to attend the sessions. Clark insisted that Aspen no longer permits members to bring family, but in February two members, Republican Jan Schakowsky of Chicago and Zack Wamp of Tennessee, brought their children to Aspen's "No Child Left Behind Conference" in San Juan. Perhaps they misinterpreted the conference title.

As for members who tack on vacations before or after the programs, Clark said that Aspen has no control over that. Aspen provides members with full price coach tickets and they are free to adjust their arrival and departure dates, presumably so they can use their frequent flyer miles—nicely inflated by Aspen, of course—to upgrade to first or business class.

It's time to end the whole practice. Let them go to conferences in Washington.

ACTION AGENDA

If you don't want your senators or representatives spending their entire lives traveling the world for free—if you'd prefer them to focus on *your* needs instead of their own and those of the top lobbyists—you can start by letting them know you don't approve of their globe-trotting.

Three of the websites we mentioned earlier—politicalmoneyline.org, americanradioworks.publicradio.org/features/congtravel/, and citizen .com—are doing a major public service by keeping tabs on travel by members of Congress. Now, all of this information—which many members of Congress might not want you to see—is readily available to anyone who cares to look.

In the past, it was very difficult to find out who went on these trips, who paid for them, and how much they cost. It's only recently that public interest organizations have begun to expose these excesses. When it comes to "official" government-paid travel, which is also rampant, there are some subscription-based services that provide details about government-funded travel by members of the House, but unless you're willing to pay a few thousand dollars, it's still top secret. And if you're looking to see where your tax dollars sent members of the Senate, be prepared to take a trip to Washington and head for a small room where you can look at each of the reports and pay for copies. That's what passes for open government in the Senate.

KEEPING IT IN THE FAMILY

From the very beginnings of our Republic, the men and women who have held public office have had to struggle with the conflict between their desire for wealth and the constraints on public officials' salaries. Their offices give them fame and power, but often seems to make it harder to amass wealth and live the lifestyle they'd like to lead.

In the old days, public officials solved the problem by flat-out stealing and bribery. In the nineteenth century and the first half of the twentieth, small-scale corruption was more the norm than the exception. When Harry Truman was elected to the United States Senate in 1934, even he added his wife, Bess Truman, to his taxpayer-funded Senate payroll, at a salary of $2,400 a year (senators themselves made $8,500 at the time). Her salary was much higher than that of any other employee in his Senate office.[28] It appears that she had a no-show job, which was apparently not uncommon during that period. Nevertheless, according to Truman biographer David McCullough, Truman grew worried that there might be adverse publicity about the exact nature of her position; he took a preemptive step, suggesting that his wife "only just drop in and do some signing of letters . . . it helps all concerned."[29]

As Senate majority leader, Lyndon B. Johnson used his power to preserve a monopoly for the radio station he owned in Austin, Texas. As long as the larceny wasn't too blatant, it was winked at and tolerated.

Back then, congressmen and senators could do what they liked with campaign funds (there was no public disclosure requirement) and could raise as much money as they needed from whomever they wanted (since there were no contribution limits). Cash was often at the core of American politics. If a candidate pocketed some of the campaign contributions and spent them on personal things, there was no way to find out.

All that began to change in 1971, when campaign reform legislation capped donations and required disclosure of all contributions. Several years later, after the Watergate scandal of the early 1970s, everything changed. Laws that had long sat neglected on the statute books were dusted off and enforced. Suddenly, cash was a no-no—now, politics had to be conducted on a checkbook basis. Expenditures and donations had to be disclosed, their amounts and purposes strictly limited.

Many of the Old Bulls of Congress continued to do business as usual, trying to circumvent the new rules. House baron Dan Rostenkowski (D-IL) went to jail for using his office budget to buy postage stamps and cashing them in for his personal use. House Speaker Jim Wright (D-TX) arranged for a friendly donor to buy thousands of his books so he could increase his royalties. He was ousted from office over the scandal. By then, prosecutors had lost their sense of humor and many of the old-timers paid for such creativity with their careers—and even their liberty.

But the conundrum remained: how do you match the power and fame of public office with money sufficient to feed the desire for omnipotence that brings so many into public life in the first place? Even with a Congress increasingly populated by millionaires, the vast majority of its members were still trying to eke out a two-home existence (one in Washington and one in the district) on $165,200 a year. Senator Warren Rudman (R-NH), one of the best, once told us: "When I came to Washington, I had a large Merrill Lynch Cash Management Account. Now it's all gone." One senator who was too honest to steal, he left after two terms "for financial reasons."

Today, sanctioned nepotism is a thing of the past in the House or Senate offices. Yet some members have recently found a new way to channel money legally to their family members.

Although members of Congress are now explicitly prohibited from hiring family members to work on their official staffs (except for spouses who served on staff before marriage), they are allowed to employ family members to work on their campaign and political action committees—as long as they're paid "no more than the market value for bona fide campaign services." [30]

The spouse-hiring floodgates opened in 2001, when the Federal Elections Commission (FEC) issued a formal advisory opinion allowing this exception for campaign staffs. The ruling was in response to a request from Congressman Jesse Jackson Jr. (D-IL), who sought guidance about whether he could hire his wife Sandi to work on fund-raising and administration for his campaign. In Jackson's case, the FEC considered Sandi Jackson's background and experience in working on political campaigns and determined that hiring her was appropriate. In finding her hiring to be permissible, the FEC indicated that the contract for her services must contain the language that is customarily used between campaign committees and consultants.

The FEC also ruled that any payment to a family member that was in excess of the fair market value of the services would be considered to be a "personal use of campaign funds."[31]

The FEC's positive ruling on Congressman Jackson's inquiry about hiring his wife apparently gave other members the same brilliant idea. All of a sudden, dozens of congressmen began to put their wives on their campaign or political action committee payrolls. Most claimed that their spouses had been doing the same work as volunteers for years.

But regardless of whether such a tactic is legal—which it shouldn't be—the practice of hiring family members raises obvious ethical questions: does the person hired have the background and experience to make a legitimate contribution to the campaign? Or is the family member simply being placed on the campaign or PAC payroll as a way to add income to the member's family? And, are some of these jobs really no-show arrangements? Is the spouse in question being hired as a way of circumventing the congressional prohibition on outside income limitation? Is the payment merely a conversion of federal campaign funds for personal use?

The opportunities for abuse are apparent, but it's very difficult to identify and document violations. For the most part, the information about such payments isn't easy to find. To determine whether a member of Congress is paying a family member from campaign funds, one would first have to search each and every financial disclosure filed with the FEC by every campaign committee and every political action committee. Assuming that a payment was made directly to the spouse, child, or brother in his or her own name, one might find it. But many such family members set up their own companies, and their business names provide no clue to their relationship with the member. For example, Julie Doolittle, the wife of California congressman John Doolittle, was hired to do fund-raising for his political action committee. She formed a company named Sierra Dominion Financial Solutions—a generic-sounding moniker that would be almost invisible to anyone examining the congressman's records to look for payments to his wife.

Even if a family member can be identified and the transaction scrutinized, though, anyone trying to evaluate such arrangements also has to make a subjective assessment of the character of the work done, and the experience and abilities of the family member performing the work. Those are not easy tasks.

Most members of Congress don't hire their families to do anything at all. They understand that it's completely inappropriate to raise contributions that will effectively be turned over to the candidate's family for personal use. But it turns out that a lot of members of the House actually do hire their wives or children to work on either their campaigns or for their political action committees. One favorite spot for congressional wives is in fund-raising. Many of the congressmen who hire their wives to ostensibly raise campaign money arrange to pay them a percentage of the total amount raised. What's wrong with this? Well, for starters, most of the candidates who do this are well-entrenched incumbents with prime committee assignments who have no trouble attracting donors. They don't need serious fund-raisers. Everyone knows they're open for business; after years or decades in office, their funding networks are well oiled. And, in a number of cases, the wife or child has no experience in campaign work or fund-raising.

But the worst part of this practice is that it potentially transforms every campaign contribution into a personal bribe. The campaign money a congressman raises stays in his campaign account, and his ability to use those funds is strictly regulated and limited. But the minute his wife is on the payroll, the money she is paid presumably goes into their joint checking or savings account, and is part of the family income. So when a PAC or donor writes a check to a congressman's campaign committee, isn't it really a direct payment to the congressman through his wife—even if the wife is a talented fund-raiser, which generally isn't the case?

There is one instance where no one would claim that a certain congressman's wife serving on his campaign payroll lacks the appropriate experience. Rep. Dana Rohrabacher (D-CA) pays his wife, Rhonda Carmony, to manage his House campaigns. What's her experience? Well, in 1996 she was accused of dirty tricks in a California state race. Ultimately, she pled guilty to falsifying campaign papers for a dummy candidate and was sentenced to mandatory community service, a fine, and a prohibition on her working on any campaigns other than her husband's.[32] Would any member in his or her right mind ever hire someone who was convicted of campaign law violations and barred from all other political campaigns? Apparently only if the person was a spouse.

Former House majority leader Tom DeLay (R-TX) is a good example of another member whose wife was very much on the payroll. Does anyone seriously believe that Mrs. DeLay—who never had a paying job before she was hired by his political action committee—was out there calling and lining up donors for him? No way. DeLay was in a unique position to help the lobbyists and special interests who contributed to his campaign, and he did so. In return, campaign money came pouring in. He definitely didn't need his wife manning the phones. Yet she and her daughter were paid almost half a million dollars!

Mrs. DeLay wasn't alone. As the list below indicates, there are plenty of congressional wives who've been hired to do fund-raising and other tasks.

CONVERTING CAMPAIGN MONEY TO PERSONAL INCOME: LEGISLATORS WHO'VE PUT THEIR SPOUSES ON THE PAYROLL

- Starting in 2001, **Tom DeLay**'s wife, Christine, and his daughter Danielle were paid a total of $473,801 for work on both his campaign and his political action committees. According to a groundbreaking article by Richard Simon, Chris Neubauer, and Rone Tempest in the *Los Angeles Times,* DeLay's daughter served as a manager for his re-election campaigns and also did fund-raising for his political action committee. His wife provided "strategic guidance" to the political action committee.[33] Before she was hired for these high-level positions, Mrs. DeLay was a homemaker. Christine DeLay was also paid $3,200 per month for three years by the Alexander Strategy Group, the lobbying firm of Edward Buckham, DeLay's former chief of staff. Buckham allegedly worked closely with Jack Abramoff, the convicted lobbyist, on behalf of Russian oil and gas interests who wanted DeLay's support on specific legislation.

  What were Christine DeLay's responsibilities in her job for the lobbying firm? To provide lists of every congressman's favorite charity. That's certainly worth $38,000 a year, isn't it? Buckham also set up a retirement account for Mrs. DeLay, estimated at about $25,000. The total that she earned from Buckham is estimated at roughly $115,000.[34] Buckham has been described as a "decision maker" for Americans for a Republican Majority (ARMPAC), which paid

DeLay's wife and daughter—and his daughter's Texas business—$350,304.[35]

- **Rep. Richard Pombo** (R-CA) paid his wife and two brothers $357,325 from 2000 to 2004 for providing bookkeeping, fund-raising, and consulting services. Prior to that, his wife, Annette Pombo, was a homemaker who had contributed "a recipe for Apple Walnut Crisscross Pie to her husband's official website."[36] In 2003, Annette began to work for the campaign; she was paid $85,275 for her work.[37] Pombo's brother Randy served as campaign manager and treasurer; he was paid $272,050 starting in 2002. According to the *Los Angeles Times* article, another brother, Ray, "sometimes catered the congressman's tri-tip and oysters barbecue fund-raisers."[38] As the *Times* reporters note, "In the 2003–04 campaign cycle, Pombo paid more to his family members—$217,000—than his opponent, Jerry McNerney, spent on his campaign. McNerney, a Pleasanton mathematician, spent $154,677. He lost to Pombo 61 percent to 39 percent."[39]

- **Rep. Zoe Lofgren** (D-CA) paid her husband's law firm $251,853. She claims that the firm also works for other politicians.[40]

- **Rep. Howard Berman** (D-CA) paid his brother Michael's political consulting firm $205,500 from 1998 to 2002.[41]

- Then representative, now senator, **Bernie Sanders** (I-VT) paid his wife, Jane O'Meara Sanders, $91,020 between 2002 and 2004 for consulting and media time buying for his campaign. He also paid his stepdaughter Carina Driscoll $65,002 as "wages" from 2002 to 2004. His stepdaughter is a former state legislator who had previous experience running Sanders' earlier campaign.[42]

- Julie Doolittle, wife of **Rep. John T. Doolittle** (R-CA), was paid more than $160,000 from Rep. Doolittle's leadership PAC starting in 2003. Mrs. Doolittle's creatively named firm, Sierra Dominion Financial Solutions, apparently operates from her home; it has no other employees, no website, and no phone listing. She takes a 15 percent commission from the PAC contributions she raises.[43] Mrs. Doolittle was also hired to do charitable fund-raising by convicted lobbyist Jack Abramoff and his firm Greenberg Traurig.[44]

NOTE: Congressman Doolittle, much to his credit, announced in January 2007 that his wife will no longer work on his campaign committee staff. Good for him! But get this: Doolittle now claims he owes his wife's company $139,000 from his last campaign—almost $100,000 of which was incurred after the election, according to Doolittle's recent FEC filing.[45]

- **Sen. Barbara Boxer** (D-CA) paid her lobbyist son, Douglas, $130,000 over a four-year period to manage her political action committee.[46]

- **Rep. Ralph Hall** (R-TX) paid his daughter-in-law $123,761 for campaign work.[47]

- **Rep. Pete Stark** (D-CA) hired his wife, Deborah Stark, as his campaign manager; he has paid her more than $119,000 since 2000.[48]

- **Rep. Howard "Buck" McKeon** (R-CA) paid his wife $152,363 over a four-year period; before that, she worked for free.[49]

- **Rep. Jerry Lewis** (R-CA) retained his wife, Arlene Willis, as chief of staff of his congressional office at a salary of nearly $111,000 a year after their marriage. It is perfectly legal to have a spouse on your congressional staff if the person had the job before the marriage. But that's not all: Lewis's stepdaughter, Julia Willis-Leon, who once owned a wedding company in Las Vegas,[50] received $44,474 from the Small Biz Tech PAC for commissions and reimbursements for fundraising for the PAC. Lobbyist Letitia White, a former Lewis aide, was the largest contributor to the PAC, which gave away only five thousand dollars or so to congressional candidates. White is a lobbyist for Trident Systems, a defense contracting company that received more than $1 million in earmarked funds in 2006.[51]

- **Rep. Ron Lewis** (R-KY) hired his wife, Kayl, as his campaign manager for $50,000 per year.[52]

- **Rep. Bart Stupak** (D-MI) has paid his wife, Laurie, about $36,000 annually for the past two years as finance director for her husband's campaign. Last November, she received a bonus of $2,500.[53]

- **Sen. Joe Lieberman** (D-CT) paid his son Matthew about $34,000 and daughter Rebecca about $36,000 for working on the senator's 2004 presidential campaign.[54]

- **Rep. Jim Costa** (D-CA) hired his cousin, Ken Costa, as his "co-campaign manager" and paid him about $45,000 in 2004.[55]

- **Rep. Dana Rohrabacher** (R-CA) paid his wife, Rhonda Carmony, $40,000 a year for her work as his campaign manager. From 2000 to 2004, she received $114,804.[56] When Rohrabacher and his wife were expecting triplets in the spring of 2005, convicted lobbyist Jack Abramoff and former congressman turned lobbyist David McIntosh and their wives invited friends of the couple to a baby shower.[57] Guests were invited to contribute to a fund for a DC nanny service. A total of three thousand dollars was raised. Rohrabacher made no disclosure of the gift, since he had asked for and received a waiver of the reporting requirement of any gift of more than fifty dollars.[58]

- **Rep. Bob Ney** (R-OH) paid his wife, Elizabeth, about $1,730 a month during his 2004 campaign. She has worked as a campaign consultant for him since the 2001 election cycle.[59]

- **Rep. Dave Reichert** (R-WA) hired his nephew, Todd Reichert, as driver for three thousand dollars last year and threw in several hundred dollars for mileage.[60]

- **Rep. Chris Cannon** (R-UT) hired all three of his college-age daughters to work on his last campaign. Emily was paid $5,425, Jane $9,508, and Laura $17,766.[61]

- **Rep. Lincoln Davis** (D-TN) hired his sister-in-law Sharon Davis as his campaign treasurer in 1994, paying her one thousand dollars per month. More recently, in the last half of 2004, the representative hired his daughter Libby Davis as a campaign coordinator for $2,334 a month.[62]

- **Rep. Louie Gohmert** (R-TX) employs his wife, Kathy, as his campaign manager. She was paid $21,791 in a four-month period, including a $7,500 bonus last November.[63]

- **Rep. Tim Bishop** (D-NY) hired his daughter Molly, paying her $46,995 as the finance director of his 2004 campaign.[64]

- **Rep. Bob Filner** (D-CA) paid his wife $505,000 for her fund-raising skills during the ten-year period from 1995–2005.[65] Mrs. Filner has not registered her business in D.C., has no business telephone listing, and apparently has only one client—her husband. Even when her husband was unopposed, as in 1998, Mrs. Filner was still paid $60,000 for her services. Although Congressman Filner insists that his wife is an experienced and competent consultant, he refused to let her be interviewed by a local newspaper, claiming that no employees or vendors are permitted to make statements to the press.

- **Rep. J. D. Hayworth** (R-AZ) paid his wife, Mary, $20,000 a year to run his political action committee. She is his only employee.[66] He was defeated for reelection in 2006.

- **Rep. Elton Gallegly** (R-CA) pays his wife, Janice, two thousand dollars a month to work on his fund-raising. Before that she had worked for free for eighteen years, but recently decided that she wanted to be paid so that she had "financial independence." Her husband approved: "I think that it's important that she have a little more independence and not feel like she has to depend on me if she needs a couple hundred dollars or if she wants to buy something," he said.[67]

- **Rep. Scott McInnis** (R-CO) paid his wife, Laurie, $39,000 for consulting services—after he decided not to run again.[68]

- **Rep. John Sweeney** (R-NY) paid his wife, Gayle Ford, a 10 percent commission on all funds raised for his campaign. She has received $72,849 since 2003.[69]

- **Rep. Jeff Flake** (R-AZ) hired his wife, Cheryl, to work on his campaign. She has been paid $27,000 since the beginning of 2001. Flake also hired his brother-in-law and paid him a $20,000 fee.[70]

- **Rep. Ed Pastor** (D-AZ) temporarily hired his nephew for a thousand dollars a month during a five-month period "to fill in for a hospitalized staff member."[71]

- **Rep. John Shadagg** (R-AZ) hired his son for $915 for "data entries" and "installing campaign signs."[72]

- As noted, **Rep. Jesse Jackson Jr.** paid his wife, Sandi, $60,000 for fund-raising from January 2005 to June 2006.

So how can we tell if a representative is simply hiring his or her spouse as a way to bring in income through the back door? Here's a checklist of possibly troublesome hirings of congressional family members.

ALL IN THE FAMILY:

How to Tell If Your Representative's Family Hire Is Legitimate or Ludicrous

1. Is the family member hired and paid in his or her own name?
2. Does the family member have relevant background and experience for the job?
3. If the person operates as a business or corporation, does it have any other clients?
4. Does the business have a telephone listing in its own name?
5. Is the salary commensurate with what comparable professionals would make?

6. Is the work that the family member is hired to do appropriate for the campaign?

7. Did the candidate and family member make an open and honest disclosure of the work and their relationship?

ACTION AGENDA

It's time for Congress to change the rules and prohibit payments to any family member by a congressional campaign committee or a political action committee. It looks bad and it undermines our confidence in the electoral system.

Even though only two senators have recently hired family members on their campaign staffs —Sen. Barbara Boxer (D-CA) and, during the 2000 presidential campaign, Sen. Joe Lieberman (D-CT)—the newly Democratic Senate has killed a bill that would have banned payments to family members from PACs and campaign committees. By 54 to 41, the Senate voted to table the amendment on January 10, 2007. Only ten Democratic senators voted in favor of the reform: Bayh, Cantwell, Feingold, Kerry, Landrieu, Mikulski, Nelson, Obama, Tester, and Wyden. Presidential candidates Biden, Clinton, and Dodd did not support the reform.

Since most of the abuse in this area occurs in the House, the practice should be banned for all House members. So far, they've done nothing.

And, while we're waiting for Congress to pass this needed legislation, the House and Senate and the FEC should immediately require that each member publicly disclose any and all payments to family members in a separate document available online for the public to access. (For more on the subject, see politicalmoneyline.com, publicintegrity.org, and citizen.org.)

CONGRESSMEN AND THEIR LOBBYIST RELATIVES

More than thirty members of Congress have hired their wives to work on their campaign payrolls, as documented above. But another growing trend is that members' wives, husbands, sons, daughters, sons-in-law, or daughters-in-law are actually becoming lobbyists. That way, they get paid to lobby their spouses or parents!

It's a neat gig. Even if your family consists entirely of dullards, they're a cinch to get a job if all they have to do is lobby dear old Dad. With so many goodies on the line—through government contracts, special interest legislation and regulations, and budgetary earmarks—everyone can play the game. Usually, the legislator piously declares that his family/lobbyists could not lobby *him*. One even barred her from his office. But nobody is fooled. Everybody knows what's really going on.

It's actually been going on for a long time. Among the first to do it, in the 1970s, was Marion Javits, wife of then-New York senator Jacob Javits. Mrs. Javits became the vice president of Rudder & Finn at $67,000 a year. One of her clients was the Shah of Iran, who was seeking to promote air travel to his country. When the arrangement was revealed, a storm of negative publicity ensued, and Marion quit two weeks later.

Today, however, there are no such sensitivities. Dozens of legislators—including virtually all the leaders—have family members as lobbyists. It has gone from a source of scandal to a perk of the job.

But it's still an outrage!

The only reason these men and women are being hired as lobbyists is because of their ability to use their names and connections to influence legislative or executive action. If their names were Smith or Jones, they couldn't get near a top K Street job—unless the majority leaders were Smith or Jones.

But this blatantly corrupt practice is not against the law or contrary to congressional rules. As the *Washington Post* observes:

> The House manual says: "Neither federal law nor House rules specifically precludes a member's spouse from engaging in any activity on the ground that it could create a conflict of interest with the member's official duties." But it cautions, "the question may arise as to whether the member is improperly benefiting as a result of the spouse's employment." The ethics committee tells lawmakers "to avoid situations in which even an inference might be drawn suggesting improper action."[73]

Who are the offenders? Which legislators cash in by farming out their family members to lobbying firms? You don't have to look too far. Just start at the top. The leadership in both houses is filled with family lobbyists.

In a surprising move, the Senate has amended the Conflict of Interest laws to require senators to prohibit staff members from having any official contact with the member's spouse or immediate family if they are registered to lobby. This is not an earth-shattering reform. Even though the spouse-lobbyist can obviously talk directly to his senator family member, family members are now prohibited from using their office resources. Regrettably, the measure does not prohibit the family member from lobbying *other* senators—a necessary next step. Family members are likely to be treated with more interest and respect than other lobbyists. Many of the lobbyists' family members will also be personal friends of other members, and that itself may attract special courtesies. Furthermore, other members may want favors from the lobbyists husband/wife/father, etc. The whole arrangement stinks and needs to be cleaned up. What follows are some examples of why change is needed.

Senate Majority Leader Harry Reid: My Three Sons . . . and a Son-in-Law

The brand-new Senate majority leader has been one of the worst offenders in this department. He's been pledging to clean up Congress—let's see if he starts with his own family! Compared to other members of Congress with family members as lobbyists, the *Los Angeles Times* reported that Reid is "in a class by himself."[74]

The *New York Times* revealed that Reid "has had three sons and a son-in-law involved in firms that lobby the government or litigate for industries seeking government benefits. Hundreds of thousands of dollars have been paid for their services. This has made Mr. Reid, a member of the Senate Select Committee on Ethics, the focal point for the critics demanding that efforts be made to regulate such cozy practices."[75]

One of Reid's sons and his son-in-law have lobbied in Washington for companies, trade groups, and municipalities seeking Reid's help in the Senate. "A second son has lobbied in Nevada for some of those same interests, and a third has represented a couple of them as a litigator. In the last four years alone, their firms have collected more than $2 million in lobbying fees from special interests that were represented by the kids and helped by the senator in Washington. So pervasive are the ties among Reid, members of his family, and Nevada's leading industries and institutions that it's difficult to find a significant field in which such a relationship does not exist."[76]

At first, Reid allowed his sons—and his son-in-law, Steve Barringer—full access to his Senate office for their lobbying work. The *Los Angeles Times* reported that "in an internal memo, [Reid's chief of staff Susan] McCue said Reid's family members had lobbied his staff by 'supplying research, technical support and strategic guidance.' She described them as 'effective advocates for their clients.'"[77]

Asked about his dealings with his lobbyist-sons, Reid said "he sees no problem with lobbying by relatives, because lobbyists' activities are 'very transparent.' . . . In September 2001, Reid sent a letter to his staff telling them that he had sought guidance from the Senate Ethics Committee, on which he sits, and had been advised that there was no restriction on lobbying by a relative of a senator. He told his staff to treat his family members who were lobbyists no better or worse than any other lobbyist."[78]

Soon after the *Los Angeles Times* started asking questions, Reid decided to ban his relatives from lobbying his own office directly. But the ban, of course, does not affect other members of their firms who want to lobby Reid's office. And, according to syndicated columnist Robert Novak: "While Reid has declared that they are barred from lobbying for their powerful clients in his office, there is little doubt that they have taken advantage of their close proximity to a powerful senator."[79]

The *Times* reported that Reid's lobbyist-sons worked to promote federal land swaps, mining interests, and the University of Nevada at Reno. It also noted that the senator did not disclose to his colleagues that he was seeking these amendments on behalf of his sons' clients. After the press attention, several of Reid's sons returned to Nevada.

And the relationship is a two-way street. Lionel Sawyer & Collins, where at least one of Reid's family members has worked as a lobbyist, has donated more than $73,000 to Reid's political action committee.[80]

The cozy family relationships with the new Democratic majority don't begin or end with Harry Reid. Pennsylvania Democratic congressman John Murtha's brother Kitt Murtha is a partner at a lobbying firm for defense contractors and technology companies.[81] Murtha was Speaker Nancy Pelosi's candidate for the House majority leadership, but was defeated by Maryland Democratic representative Steny Hoyer.

Then there's former House Speaker Dennis Hastert, who abused the system by letting his son become a lobbyist.

Former House Speaker Dennis Hastert

The Speaker's son, Joshua Hastert, decided to give up his job running a music store and record label ("Seven Dead Arson") [82] and head from northern Illinois to K Street in search of gold. Only thirty-one years old, he developed a real knack for lobbying. Somehow, he was able to get things done. When über-tech company Google recently needed a Washington lobbyist on the issue of Internet neutrality, it turned to young Joshua.

Describing his career move, he told the *New York Times:* "I realized that doing consulting and government relations on the Hill took up a lot less time than running a record store and brought in a lot more money." [83] Told you he was bright!

Originally hired by Federal Legislative Associates, a firm specializing in high-tech issues, Joshua Hastert switched to Podesta Mattoon, a powerhouse lobbying firm in Washington headed by the brother of Bill Clinton's former chief of staff, John Podesta. The firm's website tells it all:

> In addition to his wide array of policy expertise, Josh has long-standing relationships with numerous offices on Capitol Hill and in the Administration, as well as a unique understanding of the legislative process. [84]

One of those "long-standing" relationships, of course, started a long time ago—at Joshua Hastert's birth. Once he came to Washington, it proved a gold mine. Podesta Mattoon, as it happens, represents defense contractors, drug companies, Altria (the former Phillip Morris), high-tech companies, and airlines.

Although Joshua Hastert insists that he and his father never discuss his clients, they seem to be on the same wavelength. Last year, the son represented Chiron on a lobbying team on pandemic vaccines. [85] Chiron and other companies wanted immunity from lawsuits in the event of injuries or death caused by the vaccines during an epidemic such as bird flu. Coincidentally, Speaker Hastert inserted language to do just that in a defense appropriations bill—without even giving members an opportunity to debate the measure. The legislation was reportedly worth billions to the industry. In another coincidence, the pharmaceutical industry donated more than sixty-five thousand dollars to Hastert for the 2006 election.[86]

Now that there's a new speaker, it will be interesting to see if Hastert retains his clients.

Roy Blunt, the Republican Whip

Roy Blunt, the Republican Whip, has lots of lobbyists in his family. His wife, Abigail Perlman, and son, Andrew Blunt, have both lobbied for Altria, the new sanitized name for the company that owns Philip Morris. (Where did that name come from, by the way? Could they possibly be trying to get us to think of the tobacco company as "altruistic"?) Blunt married Perlman, a tobacco lobbyist at the time, on November 13, 2002, after he divorced his first wife of thirty-one years.

The new Mrs. Blunt is now a director of federal government affairs for Altria, which spent almost $7 million on lobbying last year. Before their marriage and while he was dating Ms. Perlman, Blunt tried to do her and her bosses a big favor. He quietly drafted language to benefit Altria/Philip Morris and tried to sneak it in the bill that established the Department of Homeland Security—without alerting the Republican leadership. Blunt's mission was to minimize the sale of cigarettes on the Internet, a thorn in the side of Altria/Philip Morris profits.[87] Unfortunately for the tobacco folks, the speaker of the House, Dennis Hastert, found out about Blunt's concern for the biggest tobacco company in the world and killed the amendment. As director of government affairs, Mrs. Blunt says that she no longer lobbies Congress for tobacco, focusing instead on Altria's other products.

Besides his friendship with the attractive former tobacco advocate, Blunt had other motivations for his stealth attempt to help out his industry friends. Just several weeks earlier, employees of Altria/Philip Morris, Kraft Foods, and Miller Brewing subsidiaries made thirty-seven contributions totaling $30,900 to Blunt's leadership PAC. According to Public Citizen's Clean Up Washington Project, "thirty-one of those contributions were made on October 25, 2002, the same day on which the PAC also received the then maximum $1,000 each from three Altria lobbyists: Lindsay Hooper,

Walter Steward, and Franklin Polk."[88] Some coincidence! All in all, Altria has given more than $270,000 to Blunt's political action committees.[89]

But the congressman and his wife are not the only fans of Altria/Philip Morris in the family. Andrew Blunt, the congressman's son, is also a lobbyist for the company in Missouri. A few years after graduating from law school, Andrew was singled out by the tobacco company to represent its interests, and those of its subsidiary Kraft Foods, in his home state. He also represents UPS, Burlington Northern, and SBC Missouri—all big contributors to his father.[90]

But we can't attribute all of Andrew's success to his dad's intervention. His career isn't hurt by the fact that his brother, Matt Blunt, is the governor of Missouri. Not surprisingly, Altria's top executive, John Scruggs, contributed to Matt's campaign, as did other Washington lobbyists and PACs.

Roy Blunt helped out FedEx and UPS, and both companies contributed to his PAC several months later. Blunt sponsored an amendment to the Pentagon appropriation bill to preclude companies with more than 25 percent foreign ownership from receiving air cargo contracts in Iraq. The winners: UPS and FedEx. The loser: DHL. According to the *Washington Post*, UPS was one of Andrew Blunt's lobbying clients in Missouri.

Even Blunt's daughter has recently joined the family profession. Registered as a lobbyist in Missouri, she's likely to benefit from her father's and brother's respective positions. She hasn't filed any disclosure forms yet, so it's too soon to tell who her clients will be.

Former Senate Majority Leader Tom Daschle

Though he's been purged from the Senate by the wise voters of South Dakota, former majority leader Tom Daschle helped his wife establish a lucrative lobbying practice at Baker, Donelson, Bearman & Caldwell while he was in office. According to Arianna Huffington, the firm got "$1.5 million from clients, including American Airlines, Northwest Airlines, the American Association of Airport Executives, the Cleveland airport, Boeing, Loral, and L-3 International, one of the companies chosen by Congress to supply airports with bomb detection equipment."[91]

Perhaps Tom lost his seat because South Dakota voters didn't buy Mrs. Daschle's cute explanation: "When clients retain me, they don't retain me because I am Tom Daschle's wife." Besides, she noted, the staff members she frequently lobbied "are pretty junior and may or may not know who I am."[92]

Right. That reminds us of how Hillary Clinton said in her memoirs that she didn't change her last name when she got married so that she could appear in court without anyone knowing that her husband was the Arkansas attorney general.

Just how dumb do these people think we are?

Unfortunately, the list of lobbyist-family members is quite extensive. We've listed them in alphabetical order so you can find how your local senator or congressman is serving interests other than yours! The list includes twenty senators—one-fifth of the body—whose family members have chosen lobbying as their career!

CASHING IN: PUTTING THE WIFE AND KIDS TO WORK LOBBYING YOU!

- **Sen. Evan Bayh** (D-IN): his father, former senator Birch Bayh, is a lobbyist for various telecommunications and other corporations
- **Sen. Jeff Bingaman** (D-NM): his wife, Anne Bingaman, is a lobbyist who used to work for Global Crossings[93]
- **Sen. Barbara Boxer** (D-CA): her son Douglas is a lobbyist for tribal interests and airports
- **Andrew Card Jr.:** the former White House chief of staff has a brother and sister-in-law, Brad and Lorine Card, who are lobbyists
- **Rep. William Clay** (D-MO): his daughter represents the city of St. Louis as a lobbyist
- **Sen. Kent Conrad** (D-ND): his wife, Lucy Calautti, is a lobbyist for Major League Baseball
- **Rep. Christopher Cox** (R-CA): his wife, Rebecca, is a lobbyist for Continental Airlines
- **Rep. William Delahunt** (D-MA): his daughter, Kara Delahunt, is a lobbyist for the government of Colombia
- **Rep Tom DeLay** (R-TX): he is no longer in Congress, but when he was, his brother Randolf lobbied for Reliant Energy

- **Sen. Elizabeth Dole** (R-NC): her husband, Bob Dole, has been a lobbyist for a variety of health care and other corporations, including Tyco International

- **Sen. Byron Dorgan** (D-ND): his wife, Kim, is a lobbyist for the American Council for Life Insurance. But don't jump to conclusions: a spokesman for the Council said that it had hired her because of her lengthy lobbying experience rather than her family ties

- **Sen. Dick Durbin** (D-IL): his wife, Loretta, is a lobbyist, but, in expiation, she represents the American Lung Association, hardly a special interest group

- **Sen. Michael Enzi** (R-WY): his son, Brad, is a lobbyist

- **Rep. Harold Ford Jr.** (D-TN): the former congressman's father, also a former congressman, has been a lobbyist for Fannie Mae and a number of health care organizations. Harold Jr. just lost a bid to become senator from Tennessee and is now heading the Democratic Leadership Council.

- **Sen. Tom Harkin** (D-IA): his wife, Ruth, is a lobbyist for United Technologies, which gets huge defense contracts while Senator Tom sits on the Appropriations Committee[94]

- **Sen. Orrin Hatch** (R-UT): his son Scott D. Hatch lobbied for Walker, Martin, & Hatch. In 2003 he represented the manufacturers of ephedra while his father sponsored legislation to exempt the diet supplement business from federal regulation. The *Los Angeles Times* reported that Scott got more than $2 million in lobbying fees. The FDA banned ephedra after dozens of deaths. But not before Senator Hatch got more than $137,000 from the diet supplement industry. The senator's other son, Parry, lobbies for the National Nutritional Foods Association[95]

- **Sen. Mary Landrieu** (D-LA): her aunt Phyllis Landrieu, a lobbyist for Tenet Healthcare, was paid $150,000 in 2005. Mary Landrieu herself won office as the daughter of Moon Landrieu, mayor of New Orleans

- **Sen. Pat Leahy** (D-VT): his son, Kevin Leahy, is a lobbyist

- **Sen. Trent Lott** (R-MS): his son, Chester T. "Chet" Lott Jr., is a lobbyist for Bell South and Edison and for the Thoroughbred Racing Association. Lott is the new Senate minority whip

- **Sen. Richard Lugar** (R-IN): his son, David Lugar, is a lobbyist for the the American Petroleum Institute, Verizon, Microsoft, Price Waterhouse

(continued)

Coopers, Tyson Foods, Natl. Hockey League, Hewlett-Packard, SBC Communications, hospitals, nursing homes, the city of Hartford, the city of San Juan, etc., etc., etc.

- **Rep. Buck McKeon** (R-CA): his wife, Pat, is a lobbyist for AT&T [96]
- **Rep. John Mica** (R-FL): his son, Dan Mica, lobbies for the Association of Credit Unions
- **Rep. George Miller III** (D-CA): his son, George Miller IV, lobbies in Congress for a home building company, but in California his firm was paid $300,000 by Phillip Morris in 2005. He also represents Intuit, HSBC, and other corporate interests
- **Rep. David Obey** (D-WI): his son Craig is a lobbyist, albeit for the National Parks Conservation Association
- **Rep. Nick Rahall** (D-WV): his sister Tanya Rahall is a lobbyist for the government of Qatar. She's paid $15,000 per month
- **Sen. Pat Roberts** (R-KS): his son David is a lobbyist.
- **Sen. Debbie Stabenow** (D-MI): her son Todd is a lobbyist for the Michigan Credit Union League
- **Sen. Ted Stevens** (R-AK): his wife, Catherine Stevens, is a lobbyist for Mayer, Brown, Rowe & Maw. Stevens is also dodging bullets after a *Los Angeles Times* story reported that he tried to save a $450 million defense contract for a local businessman who had previously been his partner in a lucrative real estate deal [97]
- **Sen. John Sununu** (R-NH): his father, John, President George H. W. Bush's chief of staff, lobbies for the U.S. Chamber Alliance for Energy and Economic Growth
- **Sen. George Voinovich** (R-OH): his son George F. Voinovich is a lobbyist [98]
- **Rep. Curt Weldon** (R-PA): his daughter, Karen, is a lobbyist; among the twenty-nine-year-old's clients, according to the *Los Angeles Times,* are Dragomir and Bogoljub Karic, Serbs with close family ties to former Serbian dictator and war criminal Slobodan Milosevic.[99] Weldon himself, fortunately, was defeated for reelection

Source: Public Citizen[100] and opensecrets.org

The list of self-serving sons, daughters, wives, husbands, sisters, brothers, and fathers of legislators who work the other—and more profitable—side of the fence while their relatives serve in elected office is appalling. It raises serious questions of who these elected officials really serve: us or their own families.

The ridiculous claims these legislators make—that they're not influenced by their family, or that they treat their sons and daughters like any other lobbyist—do not deserve to be treated at face value. Who are they kidding?

ACTION AGENDA

The House and the Senate can set rules for its members. Both bodies should refuse to allow any spouse or direct relative of any members to serve as lobbyists or to be principals in any firm that engages in lobbying. We must demand this ethical change of our Congress.

In the near term, if you live in a district or state with a congressman or senator whose family lobbies, vote against him and send him a message. He can't serve them and you at the same time!

MORE CONGRESSIONAL FAMILY VALUES, WASHINGTON STYLE

Wanda, Washington's Wackiest Wife
Montana Senator's Wife Physically Attacks Woman in Garden Center Because She Had to Wait to Get Mulch

There is a disease that inflicts people who have been political insiders in Washington, D.C., for too long. The main symptom, seen all too often, is that they feel that they are entitled to be treated like royalty, simply because they're elected officials—or married to them.

Here's a revealing story about one of them:

In May 2004, Wanda Baucus, the wife of Senator Max Baucus (D-MT), was arrested for assaulting a woman in a garden center in the Washington,

(continued)

D.C., area. Apparently, Mrs. Baucus was annoyed because she had to wait while the garden center staff helped another, rather ordinary, woman to her car with heavy bags of mulch before they helped her, a senator's wife.

Enraged by this outrageous treatment of an obvious VIP, Mrs. Baucus used a large bag of mulch and her silver BMW (while her bichon frise was sitting in it) to block the woman's car and prevent her from leaving the parking lot. When the other woman tried to leave, an argument broke out. According to witnesses, Mrs. Baucus "pummeled," scratched, and hit the other woman, causing bruises to her face and scratches on her shoulder.[101] She then left the scene.

According to the local NBC report, court records indicate that Mrs. Baucus then called the police and claimed that she had been the one who was assaulted.[102] But evidently the police did not believe Mrs. Baucus after they interviewed witnesses, weighed her own "conflicting"[103] stories, and observed that the victim had a "swollen left cheek bone and scratches and redness on her left shoulder blade."[104]

Mrs. Baucus later returned to the garden center with her husband, the senator, to try and spin her story to the police who had been called to the scene. It didn't work.

The next morning, Senator Baucus released the following statement:

> There was a situation with Wanda last night. We're trying to sort it out, going through the proper channels. I stand by her 110 percent, and she has my full support.[105]

Situation? A *situation* with Wanda last night? Hey Max, here's the "situation": your wife punched and scratched a woman because she didn't feel that a United States senator's wife should have to wait a few minutes for her mulch to be loaded into her car. She was furious that a mere average citizen was given assistance ahead of her. And then she lied about it, trying to blame everything on the victim. But no one believed her because witnesses—and the condition of the woman's face—left no doubt that Wanda was the aggressor.

The day after the "situation," a warrant for Wanda's arrest on misdemeanor assault charges was issued. She appeared in court and was released on her own recognizance—on the condition that she stay away from the victim and the garden center.

Did Senator Baucus really view an unprovoked physical attack over a bag of mulch as a "situation"? A "violent tantrum" is more like it.

And he "stands by her 110 percent"? Have these people ever heard the word *apology?* There is no record of any apology by either Senator or Mrs. Baucus for her wacko behavior. Instead, she vehemently denied the stories that were printed about the incident and told the *Washington Post* that "the truth will come out." [106] And it does not seem that this experience, one that would humiliate and sober most people, was at all upsetting to Wanda. According to *The Hill,* Mrs. Baucus seemed "oddly carefree" when interviewed three days before her court appearance. [107]

Eventually, Mrs. Baucus worked out a deal with prosecutors that required forty hours of community service. We can only hope her service will include some anger management classes.

On one level, of course, Wanda Baucus should be relieved: the truth did come out, after all. It's just that the truth is that Wanda was an angry and violent woman that day, who thought she deserved special treatment just because her husband is a senator. It's a disease that is caused by being a Washington insider for too long.

The garden center assault wasn't Mrs. Baucus's only bizarre episode. One former neighbor, an Italian journalist named Beppe Severgnini, described her erratic behavior in 1995. On the day that he was moving, Severgnini recalled, Wanda Baucus came out on the street and yelled at him to get his moving van out of the way. "Get away from there right away," she screamed. "I'm a senator's wife." [108] Severgnini wrote about the incident in his 2002 book *Ciao, America!* two years before Wanda's assault arrest:

She was very sweet, but on the day that we had the removal van, she went berserk. I thought that it was un-American, un-Washington, un-Democratic, and in bad taste. It was simply a removal van and she was really very aggressive, shouting "I'm a senator's wife." [109]

Mrs. Baucus denied the incident—just as she denied the garden center incident.

Now get this: Mrs. Baucus's husband cosponsored a Senate resolution that designated September 26 as "Good Neighbor Day." Maybe he should have consulted his wife about the finer points of the bill.

But Wanda Baucus isn't always violent and aggressive; she has a pacifist side, too—and she wants people to know about it. In 2003, while her husband

(continued)

publicly called for the removal of Saddam Hussein, she installed an antiwar poster in the window of her Georgetown home that announced "Peace Is Patriotic" above a flag with blue doves in place of stars. When asked about it, she told the *Washington Post* that she wanted peace for all of the people in Iraq— and even for the "camels that are out grazing."[110] Mrs. Baucus told the *Post* that television images of the bombings in Iraq were keeping her from sleeping at night. "A billion Muslims all over the world are in pain to see their brothers losing their homes and their families losing the stability of their civilization." And Saddam? Well, Mrs. Baucus considered him "very proud of the history of his country." She also claimed that the United States was "way out of line" in making a preemptive strike and trying to "assassinate its leader."[111]

Thanks for your insights, Mrs. Baucus.

EARMARKING: HOW TO WASTE TAXPAYERS' MONEY . . . AND GET CAMPAIGN CONTRIBUTIONS FOR DOING IT

At last report, the federal budget deficit was $300 billion and the national debt was $8.5 trillion. Who is to blame? The list of suspects is familiar: liberals, Democrats, bureaucrats, government agencies, Congress. But the fact is that much of the waste in government is the deliberate result of vigorous work by virtually every single member of Congress—Democrat or Republican, liberal or conservative.

When the president proposes a federal budget to Congress, even when his own party controls both houses, it's usually classified as DOA (dead on arrival). The priorities the chief executive and his cabinet assign to the various claimants for federal dollars have little to do with the eventual outcome. Instead, what emerges as the final federal budget is largely the creature of 100 senators and 435 congressmen battling, horse-trading, grubbing, maneuvering, arguing, and hustling for funds for their states and their districts.

The fastest growing category of government spending is "earmarks." The current budget contains $64 billion in earmarked appropriations for 12,852 items inserted by members of Congress to benefit special claimants among their constituents. As late as 1998, only 4,219 such projects graced our budget, costing $28 billion.[112]

Here's how it works. The federal budget appropriates a lump sum to one of its agencies—say, the Department of Interior. It used to be up to the secretary who runs the agency to decide how to allocate the money: to evaluate the various programs and projects under its auspices and decide which deserve federal funding.

But the members of Congress all have an interest in making sure that the Interior Department money goes to specific people, projects, or institutions in their districts. Some of these have merit, of course, but others are just political payoffs. So a member attaches an amendment to the appropriations bill that gives the Interior Department its funding. The amendment is called an earmark. The secretary of the Interior is then legally directed to give that particular project a certain sum of money, regardless of whether she thinks it's a necessary—or even reasonable—way to spend public funds.

Typically, the members of the House and Senate who sit on the relevant subcommittee of the Appropriations Committee that draws up the budget have the biggest claim to the cookie jar. They are usually in a position to ensure that their pet projects get funded as the spending bill goes through their subcommittee. But members of other committees also are able to lay their claim to some of the money—and *virtually every member of Congress plays this game.*

The problem isn't that this money is being spent, of course. It's that the earmarking system enables it to go to our politicians' favored projects and programs—even if when the agencies who know these projects best have already decided that they don't deserve the funds. Earmarking overrides their judgment, and puts the decision in the hands of our not-always-noble political class.

WHERE YOUR MONEY GOES: EARMARKS IN THE BUDGET

- $13.5 million for the International Fund for Ireland, which helped finance the World Toilet Summit
- And, while we're on the subject, $1 million for the Waterfree Urinal Conservation Initiative

(continued)

- $6,435,000 for wood utilization research

- $234,000 for the National Wild Turkey Federation

- $500,000 for the Sparta, North Carolina, Teapot Museum

- $250,000 for the National Cattle Congress in Waterloo, Iowa

- $100,000 for the Richard Steele Boxing Club in Henderson, Nevada (courtesy of Democratic Senate Leader Harry Reid)

- $550,000 for the Museum of Glass in Tacoma, Washington

- $100,000 for goat-meat research in Texas

- $549,000 for "Future Foods" development in Illinois

- $569,000 for "Cool Season Legume Research" in Idaho and Washington

- $63,000 for a program to combat noxious weeds in the desert Southwest

- $175,000 for obesity research in Texas

- $50 million for an indoor rainforest in Iowa

- $25 million for a fish hatchery in Montana

- $200,000 for a peanut festival in Alabama

- And we have a winner: ABC News reports that "the town of Ketchikan [Alaska] has just one main road, but now it's getting more than $200 million of your tax dollars because Rep. Don Young (R-AK) wants to replace a ferry with a bridge to the next island. And he doesn't want to build a simple bridge. He wants to build one higher than the Brooklyn Bridge and almost as long as the Golden Gate. People are calling it the "Bridge to Nowhere," because it links to an island where there is an airport but not much else. The island has no roads and is home mostly to trees"[113]

After taking over the Senate, the Democrats, led by Harry Reid, put earmarking reform on the agenda. His remedy: to require senators and congressmen to list which budgetary earmarks they put in the appropriations bills. Some want this requirement to apply to all earmarks. Others just want to limit it to the small proportion (about 2 percent) that are actually spelled out in the text of the legislation itself rather than in the accompanying reports.[114] But neither reform means much. Legislators are more than happy to take credit for earmarks; they sell well back home.

PAST VS. PRESENT: FDR VS. HRC

Contrast standard practice today with what went on one hundred years ago, when we had real statesmen instead of money-grubbing politicians in our legislative bodies.

PAST: In 1910, New York state senator Franklin Delano Roosevelt sat in his seat in the chamber in the state capitol in Albany and read the budget appropriation bill. His eye stopped at Appropriations Bill 1789, an allocation of $381.54 to improve a bridge over Wappinger's Creek. He rose to ask state senator James J. Frawley, the chairman of the Finance Committee, about the item. "Oh, that is to benefit your district," Frawley proudly told Roosevelt. "The money was not expended last year and this bill makes it available to carry on the work."

"Well," FDR replied, "I haven't heard that the money is actually needed and I'd like to have it returned to the state treasurer."

In his biography of FDR, author Ted Morgan recounts how Big Jim Sullivan, one of the Senate's bosses, told the patrician future president, "Frank, you ought to have your head examined."[115]

PRESENT: In 2005, New York's senator Hillary Rodham Clinton was inducted into the National Women's Hall of Fame, a Seneca Falls museum.

In 2006, she inserted into the Courts, Transportation and Treasury Appropriations Bill an amendment giving $800,000 to the National Women's Hall of Fame to help them to move to a new building.[116]

Why do congressmen and senators work so hard to get funding for their districts? The excuse they normally give is that they are trying to create jobs and help the local economy. But closely linked to this ostensibly worthy public purpose lurks a more sinister and relevant motivation: campaign contributions!

In reality, the earmarking and pork barrel process is an elaborate scheme that allows legislators to grab public money to fund programs and pork in return for the private money they collect to finance their campaigns. Using the camouflage of a professed earnest desire to help the local economy, the senators and congressmen are really using the appropriations process to line the pockets of their campaign war chests. Here's how it works:

EARMARKS: A FOUR-STEP PROGRAM FOR CASH-HUNGRY POLITICIANS

Step 1: A business or local government agency hires a lobbyist and pays him a fee.

Step 2: The lobbyist donates money to a congressman's campaign war chest.

Step 3: The congressman earmarks cash the business or government agency wants.

Step 4: The business gives the congressman campaign contributions.

In previous decades, the top seats on the Appropriations Committee and its subcommittees, where the earmarks are typically inserted, went to the oldest members, on a strict seniority basis. These elderly gentlemen tended to come from one-party districts in the old solid Democratic South, where they faced only nominal opposition at election time. Often, Republicans didn't even bother to oppose them.

But in a major "reform," enacted in 1994 when Congressman Newt Gingrich led the GOP to take control of the House of Representatives for the first time in forty years, the seniority system was laid aside in the assignment of committee seats. Indeed, future House leader Tom Delay (D-TX) got the Republican caucus to decide to put its most vulnerable members on the Appropriations Committee, thus helping them get reelected. A seat on the Committee guaranteed access to massive campaign funding—and a good chance of a return trip to Washington after the elections.[117]

The problem with earmarks is that the projects they fund are like weeds in a carefully tended garden. Generally, in the executive branch agencies, the technical career staff tasked with spending the money Congress appropriates do a pretty good job of planning what the nation needs. But no sooner do they map out their spending plans than earmarks—inserted by congressmen—make a hash of them.

Another important and insidious thing about earmarking is that it consumes so much of the average senator's or congressman's time. Desperate for campaign contributions, members of both chambers scour their states and districts for likely contributors who want something from the federal budget. Once they develop their wish lists, and get campaign contributions from the supplicants, they spend an inordinate portion of their workweeks trying to get their little earmarks passed.

It used to be that if a congressman or senator got caught sneaking through an amendment to funnel federal money to a business through a questionable project, he was considered to be a crook. No more. Now, the members who deliver the bacon to their corporate constituents are considered savvy politicians who care about their districts!

Knowing how desperate congressmen can get, the partisan leadership of the House and the Senate—when it isn't trying to gouge the budget on behalf of their own districts—use earmarks as bait to reinforce party discipline and punish those who would think for themselves. Earmarking is crowding out legitimate legislating as a congressional function. Those who bemoan the absence of statesmen and stateswomen in Congress need only look at the time the average member has to spend raising money and passing earmarks to grasp why they have no time or energy left to see the big picture.

But the biggest problem with earmarks is that they channel billions of dollars into worthless projects, thus shortchanging vitally important national priorities. For example, earmarks play havoc with our efforts to enhance our national security. The Office of Naval Research gets $2.37 billion a year to plan and develop future projects. But $590 million, about a quarter of the total, is consumed by earmarked projects that the Office actively opposes, but which Congress funds anyway.[118] Another case of floating pork!

Congressman James P. Moran: Foisting "Project M" on American Taxpayers

Typical of these projects is the so-called Project M, aptly named because its benefactor was Congressman James P. Moran Jr. (D-VA)—quite possibly America's worst congressman.

Moran's most recent disservice to America has been promoting Project M—that is, foisting it upon the taxpayer. The project features a technology for magnetic levitation that was originally "conceived as a way to keep submarine machinery quieter, was later marketed as a way to keep Navy SEALs safer in their boats, and, in the end, was examined as a possible way to protect Marines from roadside bombs," as the *Washington Post* has reported.[119] The only problem is that Project M didn't do any of these things. The Defense Department never wanted it, but Congress appropriated more and more money for it, year after year. "It kept failing to solve any problems the Navy had," said Paul M. Lowell, former chief of staff of the Office of Naval Research.[120]

In all, $37 million of taxpayer money was earmarked for this white elephant, providing Vibration & Sound Solutions Ltd., a small Alexandria defense contractor in Moran's district, with a nice living while doing nothing to make us safer.

Why did Congressman Moran push this waste of money? He says that the twenty-five jobs that depended on this funding were important to the district. But perhaps the $17,000 in campaign contributions he got from the contractor's president, Robert J. Conkling, and his wife, were equally important.[121]

Moran isn't the only one who gets earmarks for dubious defense projects inserted into appropriations bills. The Rajant Corporation in Wayne, Pennsylvania, went to its local congressman, Curt Weldon, then the vice chairman of the House Armed Services Committee, for help in getting a $2 million defense contract. It was all in the family, since John McNichol, once on Weldon's staff, was the lobbyist retained by Rajant to get the money. Between the lobbying firm and Rajant, Weldon got $9,300 in campaign contributions between 2000 and 2004. Sure enough, Rajant got its contract.[122]

Hurricane Katrina: God Made It Rain; Congress Let It Flood

In the rush to distribute earmarked dollars, real priorities often get overlooked. The most tragic example is the failure to repair the New Orleans levees that led to the city's destruction when hurricane Katrina struck. As ABC News has reported, "This disaster—including the breaking of the levees—wasn't unexpected. Regional newspapers, the Army Corps of Engineers, and the Federal Emergency Management Agency (FEMA) itself warned that a strong hurricane could have cataclysmic consequences on New Orleans and the surrounding areas. It's been reported that just before 9/11, FEMA warned that the three biggest threats to America were a terrorist attack on New York City—a massive earthquake in San Francisco, and flooding in New Orleans, if a big hurricane hit." [123] Two out of those three have since come to pass, of course . . . and in neither case were FEMA's warnings heeded.

As the *Washington Post* has noted, Louisiana's Republican senator David Vitter presented "a chilling preview of [Katrina's] rampage." The senator displayed "a computer model of a Category 4 hurricane smashing New Orleans and flooding the city under eighteen feet of water." He said, "This isn't a simulation of World War III or 'the Day After Tomorrow' or Atlantis—but one day it may be Atlantis. . . . It's not a question of if. It's a question of when." [124]

The *Post* reports that "before Hurricane Katrina breached a levee on the New Orleans Industrial Canal, the Army Corps of Engineers had already launched a $748 million construction project at that very location. But the project had nothing to do with flood control. The Corps was building a huge new lock for the canal, an effort to accommodate steadily increasing barge traffic. . . . Except that barge traffic on the canal has been steadily decreasing." [125]

Why was the money spent on the canal lock and not on flood control, despite the warnings that the New Orleans levees were a disaster waiting to happen? Largely because the state's congressional delegation got in the way, diverting the funding from levee repair to other, politically sexier projects. Rather than let the professionals at the Corps of Engineers do their thing and make the decisions as to what projects were most urgent, the politicians kept insisting they knew better and inserting earmarks for much more marginal projects, less productive for flood prevention but better for campaign contributions. According to the *Post*, "Louisiana has received far more money for [Army] Corps [of Engineers] civil works projects than any other state, about $1.9 billion." California, in second place, received only $1.4 billion—even though it has eight times Louisiana's population. [126]

Typical were the efforts of Louisiana senator Mary Landrieu to get funds for a $194 million deepening project for the Port of Iberia that, the *Post* reported, "flunked a Corps cost-benefit analysis." Blithely disregarding the experts' findings, Landrieu "tucked language into an emergency Iraq spending bill ordering the agency to redo its calculations" so her pet project could get funding.[127]

The *Post* also noted that the Army Corps of Engineers "also spends tens of millions of dollars a year dredging little-used waterways such as the Mississippi River Gulf Outlet, the Atchafalaya River, and the Red River—now known as the J. Bennett Johnson Waterway, in honor of the project's congressional godfather—for barge traffic that is less than forecast."[128]

But it was the Industrial Canal Lock project that diverted the lion's share of the funding that should have gone into levee repair. As the *Post* reported, in 1998 the Corps forecast major increases in barge traffic between the Port of New Orleans and the Mississippi River in order to justify its decision to steer money to lock construction—when barge traffic numbers had actually been falling since 1994.[129]

In this case, Louisiana's ill-fated residents, who had to race to leave their homes and businesses when Katrina caused the levees to breach, would have been better off if they had been unrepresented in Congress—and the Army Corps of Engineers had been left to follow its own data and repair the levees.

Robert K. Dawson, who oversaw the Corps of Engineers as an assistant secretary of the army in the 1980s, said that "we all should have paid more attention to the levees. But I don't recall that any of us did. . . . I never felt like the Louisiana delegation had flood control on its mind, they were focused on navigation."[130]

A former aide to former Louisiana senator J. Bennett Johnson put it graphically: "They could have built the Hoover Dam around New Orleans with the money they brought home. But they always pissed it away on politically attractive projects."[131]

Louisiana got the Feds to spend $300,000 to house former senator John Breaux's papers in a college. Presumably he needed a dry place to put them.

PORK AND TRANSPORTATION

The special preserve of earmarks and pork barrel spending is the nation's transportation budget, which includes funding for highway construction and repair as well as for mass transit. In 2005, Bush signed a $286 billion transportation appropriations bill that included 6,371 earmarks. Almost every member got to put his or her pet project in the budget, each agreeing to vote for the others' demands in return for support on his own—a process known as "logrolling."

THE SPEAKER SPEAKS OUT. . .

HOW DENNIS HASTERT MADE $2 MILLION

Did House Speaker J. Dennis Hastert deserve to be ousted from his job? Judge for yourself:

The Speaker got $207 million earmarked in the Transportation Appropriations Bill for the "Prairie Parkway," a project slated to run through Kane and Kendell counties in his home region. The Illinois Department of Transportation isn't sure the parkway is needed, and is two years into a five-year study weighing whether to improve existing roads or to build a new road despite local opposition. But Hastert lifted the decision out of the hands of the experts and made it for them—based on politics, not on merit.[132]

Did Hastert enrich himself by funding the highway? The *Washington Post* reported that he made roughly $2 million on a land sale less than six miles from the highway route. Hastert angrily denies any connection, but the timing is suspicious.[133]

In May 2005, Hastert transferred sixty-nine acres of "previously hemmed-in land from his farm to a land trust." The *Post* reports that "Hastert [then] personally intervened during House and Senate negotiations over a huge transportation . . . bill to secure . . . $152 million to help build the Prairie Parkway . . . and $55 million for an interchange 5.5 miles from his property." As the paper points out, on December 7, 2005, the land trust sold the parcels Hastert owned for "nearly $5 million. The deal netted Hastert a $2 million profit."[134]

Hastert's attorney, J. Randolph Evans, told the *Post* that the price for the land "had been locked up [by a land speculator] in 2004" and that the "price could not have risen with the news of the Prairie Parkway funding." But who knows what anticipations might have fueled the speculator's 2004 decisions about the price of the land? If he knew that the Speaker of the House wanted to get a project funded, it was a pretty safe bet that he would succeed![135]

Tammany Hall boss George Washington Plunkett, in his celebrated turn-of-the-century memoir of American political mores, described this process as "honest graft":

> There is an honest graft and I'm an example of how it works. I might sum up the whole thing by saying: "I seen my opportunity and I took 'em. Just let me explain by examples. My party's in power in the city, and its going to undertake a lot of improvements. Well. I'm tipped off say, that they're going to lay out a new park at a certain place. I see my opportunity and I take it. I go to that place and buy up all of the land that I can in the neighborhood. Then the board of this or that makes the plan public, and there is a rush to get my land, which nobody cared particular for before. Ain't it perfectly honest to charge a good price for my investment and foresight? Of course it is. Well, that's honest graft.[136]

Hastert, apparently, feels the same way—and he isn't the only one with his snout in the trough of the Transportation Appropriations Bill. The *Washington Post* has also noted that specific earmarks were inserted into that year's bill for projects such as "$2.3 million for the beautification of the Ronald Reagan Freeway in California; $6 million for graffiti elimination in New York; nearly $4 million on the National Packard Museum in Warren, Ohio, and the Henry Ford Museum in Dearborn, Michigan; $2.4 million on a Red River National Wildlife Refuge Visitor Center in Louisiana; and $1.2 million to install lighting and steps and to equip an interpretative facility at the Blue Ridge Music Center, to name a few." The newspaper noted that "critics say [that these projects] have nothing to do with improving congestion or efficiency."[137]

When Congress inserts earmarks into traditional pork-barrel areas like highway construction or public works, it distorts the process of planning and makes for irrational allocations of funding. But when it insists on spending money in more sensitive areas, it can actually divert resources from much more important priorities.

PORK AND EDUCATION

JRL Enterprises, a Louisiana company, makes software called I CAN Learn, which is supposed to help children learn mathematics. Yet, other than JRL's own extravagant claims, there is very little to suggest that I CAN Learn significantly helps anyone learn at all. A story in the *Ft. Worth Star-Telegram* found that students in the local school district, which has invested heavily in I CAN Learn, were not doing better than the statewide average in math. Teachers complained that the software often froze in the middle of a lesson, and sometimes gave the wrong answers to test questions.

I CAN Learn may not help children learn math, but its executives are very good at political arithmetic. Unable to get the kind of support they had hoped for from the experts at the federal Department of Education, they hired former Louisiana Republican congressman Robert Livingston as their lobbyist. It was a smart move: Livingston was on tap to succeed Newt Gingrich as Speaker of the House before a sex scandal derailed his career and he resigned his seat. Before leaving Congress, Livingston had earmarked $7.3 million for JRL's software, providing the company with what *Harper's* magazine describes as "virtually all of its income for 1998." After his retirement from the House, Livingston hung up a shingle as a lobbyist, and JRL became his client. Last year's appropriation bill included three earmarks for JRL's software. In all, JRL has gotten $38 million through Livingston's work, for which it has paid the former congressman almost $1 million in fees.[138]

Livingston helped get the earmarks by guiding JRL to make strategically placed campaign contributions between 1999 and 2004, totaling $81,460, of which $14,500 went to members of the Appropriations Committee. You see how it works? Between the $1 million Livingston got, the $38 million JRL received, and the $81,000 that went to members of Congress, everybody's happy...

... except for us taxpayers—and all the children who are stuck trying to learn math on that ineffective software.[139]

PORK AND OUTER SPACE

Consider the problems NASA faces in trying to fund America's space exploration program. According to *USA Today*, earmarks are so prevalent in its budget that it must "slash science, engineering, and education programs to pay for billions of dollars in congressional pet projects, most of which have little to do with the agency's mission to explore space."[140]

The price tag for politicians' pork has grown so large that NASA may have to delay the new spaceships and rockets needed to replace the space shuttles, which are slated to be retired in 2010. Instead of launching new space projects, since 2000, NASA has found itself diverting $3 billion to political pork—money it must carve out of its $16 billion a year budget—to projects like these that are way beyond its core agenda:

- Construction or renovation of dozens of museums, planetariums, and science labs for colleges

- Computers, classrooms, and lab space for colleges and schools across the United States

- A website and laboratory for the Gulf of Maine Aquarium

Sooner or later, considering the low moral fiber of so many of our distinguished representatives in Congress, earmarking was bound to lead to outright corruption. The combination of their virtually limitless ability to earmark portions of huge appropriation bills to specific projects, and their free rein to choose which groups or organizations get the earmarks, is too tempting for many politicians to resist.

GRAFT ON THE HOUSE ETHICS COMMITTEE

Congressman Alan Mollohan was the ranking Democrat on the House Ethics Committee—until he had to quit because he found himself under FBI investigation!

Punch line? He was reelected in 2006.

Mollohan's formula was simple: get federal funding earmarked for certain companies, which his friends either owned or worked for. Next, start buying real estate, with these friends picking up half the tab. Get them to ante up campaign contributions. And then, finally, watch his net worth go up and up and up.

Mollohan is currently under investigation for three such scams. Despite his having been forced to resign from the Ethics Committee, he professes

wonderment that anyone would find his real estate deals with the beneficiaries of his earmarking a conflict of interest![141]

Here are the specifics:

- Mollohan got one friend, Dale R. McBride, the CEO of FMW Com-
posite Systems—a company that manufactures equipment for NASA
and the military—earmarks worth $4.4 million ($2.1 million from
NASA and $2.3 million from the Marine Corps). He also got Moun-
tainMade, an organization to promote the wares of West Virginia ar-
tisans, $1.1 million from the federal Small Business Association.
Then McBride and his wife turned around and bought a $900,000
farm with Mollohan and his wife that they own fifty-fifty. Any connection between the earmarks and McBride having the money to pay
for half the farm? Mollohan is shocked, shocked that anyone should
question his motives: as the New York Times reported, "Mr. Mollohan
said . . . that the potential conflict of interest 'did not occur to me'
when buying the farm with Mr. McBride."[142]

- The congressman got almost $30 million in earmarks since 1999 for
the Vandalia Heritage Foundation, which redevelops dilapidated
buildings. Vandalia's director is a former Mollohan aide named Laura
Kurtz Kuhns. The congressman and his wife then bought five empty
lots with Kuhns and her husband on Bald Head Island, North Carolina, worth $2 million.[143]

- Mollohan may have been instrumental in helping the West Virginia
High Technology Foundation, which employed his friend and dis-
tant cousin Joseph Jarvis, get a $1 million subcontract from the U.S.
Energy Department. The New York Times reported that while there is
"no evidence that Mr. Mollohan . . . intervened" to help Jarvis secure
the contract, "court documents and multiple interviews show that
Mr. Jarvis was not shy about mentioning his connections to Mr. Mollohan." The Foundation hired Jarvis to "drum up business" in part
because Mollohan told a former leader of the high-tech consortium
"to work with Jarvis." Mollohan, true to form, then teamed with

Jarvis to buy twenty-seven condominiums in a fifty-two unit building in Washington, D.C. The condos, the *Times* reports, "have more
than tripled in value, to $8 million, over the decade [since Mollohan
and Jarvis bought them]." [144]

The result of these shenanigans is that Mollohan's assets, as revealed on the
intentionally vague disclosures required by Congress, rose from less than
$500,000 in assets in 2000 to at least $6.3 million in 2004. This investment
portfolio, which generated less than $80,000 in 2000, brought him between
$200,000 and $1.2 million in income in 2004. [145]

How did Mollohan get this extra money on a salary that even now
only runs $165,200 a year? The *Times* reports that he did it through creative
use of earmarking. He has earmarked an estimated $480 million for his
district since 1995, $250 million of which went to five nonprofit organiza-
tions he set up himself: [146] the West Virginia High Technology Consortium
Foundation, the Canaan Valley Institute, a water pollution control organization, the Vandalia Heritage Foundation, and the MountainMade Foundation.

Of course, the key to Mollohan's investment strategy is his ability to
keep his seat in Congress—and that, in turn, requires campaign contributions. To procure them, he returns regularly to the friends whose pockets he
has lined with his earmarks.

The *New York Times* reviewed Mollohan's campaign contribution filings
to determine how much these five organizations donated to his campaign.
The newspaper reports that "from 1997 through February 2006, top-paid
employees, board members, and contractors of the five organizations gave
at least $397,122 to Mr. Mollohan's campaign and political action commit-
tees. Thirty-eight individuals with leadership roles, including all five chief
executives—all but one of whose 2004 salaries outpaced the $98,456 na-
tional average among nonprofit leaders—contributed, often giving the
maximum allowed. At the same time, workers at companies that do business with the federally financed groups were among Mr. Mollohan's leading
contributors. Employees of TMC Technologies, which had a $50,000 con-
tract with Vandalia in 2003, have given $63,450 since 1998. Workers at Elec-
tronic Warfare Associates and Man Tech International, military contractors
that rent space from the technology consortium and whose chief executives

are on the board of the Institute for Scientific Research, combined to give $86,750." [147]

Mollohan seems to be fated for trouble as the FBI investigation continues. But how many other congressmen and senators could bear similar scrutiny? The process of earmarking virtually hands legislators opportunities for corruption.

Lobbyists are an important part of the earmarking process. More than three thousand companies, municipalities, special interests, and institutions have lobbyists chasing earmarks on their behalf. [148] Lobbyists and their PACs have given $103.1 million to members of Congress since 1998, according to Congress Watch. [149]

More of our tax money has sunk in the costly Bermuda Triangle of lobbyists, legislators, and businesspeople willing to give campaign contributions. The cash flow is simple: the lobbyists get fees from their business/clients. They in turn directly give congressmen and senators donations and encourage their clients to send checks as well. The recipients of this largesse—the public officials—spend our tax money on projects whose profits justify the campaign contributions and lobbying many times over.

The congressmen and senators say they're working to provide jobs for their districts or their state, but it's more likely that the campaign contributions are their primary concern.

See the sidebar for a list of the biggest diners at the trough—the senators and congressmen who got at least half a million dollars in campaign contributions from lobbyists. Note how bipartisan the gravy flow is: of the eighteen senators on the list, half are from each party.

SENATORS WHO RECEIVED AT LEAST $500,000 FROM LOBBYISTS [150]

Sen. Rick Santorum (R-PA)*	$1,163,560
Sen. Arlen Specter (R-PA)	$1,019,317
Sen. Harry Reid (D-NV)	$889,223

(continued)

Sen. Richard Shelby (R-AL)	$886,982
Sen. Conrad Burns (R-MT)*	$737,868
Sen. Hillary Clinton (D-NY)	$720,477
Sen. Edward Kennedy (D-MA)	$689,386
Sen. Chris Dodd (D-CT)	$666,223
Sen. Trent Lott (R-MS)	$662,632
Sen. Ted Stevens (R-AK)	$633,120
Sen. Mary Landrieu (D-LA)	$613,214
Sen. Richard Burr (R-NC)	$587,921
Sen. Maria Cantwell (D-WA)	$586,912
Sen. Chuck Grassley (R-IA)	$586,697
Sen. Kent Conrad (D-ND)	$575,707
Sen. Patty Murray (D-WA)	$516,659
Sen. George Allen (R-VA)*	$515,678

Since defeated for reelection

Things tend to be a bit more partisan in the House of Representatives, where the Republican leadership keeps a disproportionate share of the lobbyist funding. Eleven of the top recipients were Republican, only six Democrats. (Texas Republican Tom Delay recently retired from the House. But he deserves a special place in our memory because his donations eclipsed even those of Speaker Dennis Hastert: Delay got $1,322,906 in lobbyist donations before he left).[151]

MEMBERS OF THE HOUSE WHO RECEIVED AT LEAST $500,000 FROM LOBBYISTS[152]

Rep. Dennis Hastert (R-IL)*	$926,454
Rep. John Murtha (D-PA)	$869,100
Rep. Jerry Lewis (R-CA)	$819,754
Rep. Steny Hoyer (D-MD)	$780,880
Rep. John Boehner (R-OH)	$734,868

Rep. Michael Oxley (R-OH)	$711,494
Rep. Tom Davis (R-VA)	$672,738
Rep. Roy Blunt (R-MO)	$653,571
Rep. Don Young (R-AK)	$652,448
Rep. Jim Moran (D-VA)	$644,310
Rep. Jim McCrery (R-LA)	$641,271
Rep. Charles Rangel (D-NY)	$598,742
Rep. Henry Bonilla (R-TX)	$585,568
Rep. Edward J. Markey (D-MA)	$566,908
Rep. Dave Hobson (R-OH)	$549,405
Rep. Harold Rogers (R-KY)	$544,282
Rep. Dick Gephardt (D-MO)	$533,738

* Since demoted from his position as speaker of the House

PORK AND LOCAL GOVERNMENT

The growth in earmarking has generated a feeding frenzy among local governments, who are scrambling to spend taxpayer money to hire Washington lobbyists to make sure they're not left out of the action. In 2005, local governments paid lobbying firms a total of $640 million to get earmarks inserted for them in federal appropriations bills.[153] Since 1998, the number of localities and other public bodies hiring private lobbying firms to do their bidding in Washington has doubled: as of now, at least 1,421 have hired lobbyists to prowl the halls of Congress for them in search of earmarks.[154] As *Business Week* reports, "Today the halls of Congress teem with private lobbyists representing states, cities, and all manner of taxpayer-supported institutions."[155]

The magazine cites the example of Orange County in central Florida, which hired the lobbying firm of Podesta, Mattoon, headed by former Clinton chief of staff John Podesta, to lobby for it before Congress. In return for its $120,000 investment, the county got $24 million in federal aid, mostly on transportation projects. Mayor Richard T. Crotty says he's happy with the outcome: "We need to be there on the ground floor fighting for our fair share, and the best way to do that is to hire a lobbyist in Washington." [156]

Treasure Island, Florida, And Its Lucky 7,500 Residents

The city of Treasure Island, Florida, population 7,500, needed funds to repair a bridge, so it hired the lobbying firm of Alcade & Fay—a group particularly well wired with Representative C. W. (Bill) Young, chairman at the time of the House Appropriations Committee. The town had requested $15 million for the bridge repair; Young awarded them $50 million. Later, he secured them $500,000 to fix a sewer plant, $625,000 for walkways over dunes, and $450,000 for pedestrian crosswalks. For this largesse with our money, Alcalde & Fay continues to receive a $5,000 monthly fee. Mayor Mary Maloof says Alcalde & Fay are "worth every penny they get." [157]

Alcade delicately credits its ability to deliver goodies for its clients to its ability "to significantly participate in the political and fund-raising process" because it "enjoy[s] a considerable clientele from the private sector," which "might at times better enable [it] to access public policy makers on behalf of all the firm's clients." [158] In English, that means they represent a lot of fat-cat corporations and individuals whom they regularly soak for donations to legislators who, in return, are usually willing to dish up earmarks for Alcade's clients when requested to do so. For local governments, which cannot make campaign contributions, this is an important selling point indeed.

Ronald D. Utt, a senior fellow at the Heritage Foundation has a less kind description of this sort of practice. "It goes beyond mere influence peddling to just outright, classic third-world corruption." [159]

Sometimes the relationship among corporations, lawmakers, and lobbyists becomes particularly cozy. Pennsylvania's now defeated Republican senator Rick Santorum is the top recipient of campaign contributions from special interest political action committees. But coming in second was an old favorite: New York's adopted senator, Hillary Rodham Clinton.[160]

Hillary's Deal: Contributions for Favors

Hillary's I'll-scratch-your-back-if-you-scratch-mine relationship with firms like Corning Inc., an upstate New York company, shows why Clinton had received $417,575 from lobbyists and PACs by May 2006!

In a series of transactions throughout her Senate career, the *New York Times* has documented, Hillary Clinton has produced for Corning—and the company has paid up in campaign donations. As the *Times* reports, the process began in April 2003. "A month after Corning's political action committee gave $10,000 to [Hillary's] reelection campaign, Mrs. Clinton announced legislation that would provide hundreds of millions in federal aid to reduce diesel pollution, using, among other things, technology pioneered by Corning." [161]

Then, "in April 2004, Mrs. Clinton began a push to persuade the Chinese government to relax tariffs on Corning fiber optics products, inviting the Chinese ambassador to her office and personally asking President Bush for help in the matter. One month after the beginning of that ultimately successful effort, Corning's chairman, James Houghton, held a fund-raiser at his home that collected tens of thousands of dollars for her reelection campaign." [162]

In all, the newspaper reports that Corning's executives and employees have given Mrs. Clinton $137,000 in contributions since she first ran for the Senate. [163]

Senator Ted Stevens and Thad Cochran: Porker in Chief, and First Runner-Up

In any discussion of earmarking and pork, however, two senatorial "pigs"— the kings of pork—deserve special mention: Senator Ted Stevens of Alaska and Senator Thad Cochran of Mississippi. Sitting atop the appropriations process, these two Republicans have made a mockery of their avowals of fiscal conservatism. Each claims to be an opponent of big government, but between them they have accounted for close to two-thirds of a billion dollars in earmarks and pork for their states.

Stevens has gotten $325 million in earmarks in the 2006 appropriations cycle, which works out to more than $500 for every man, woman, and child in Alaska. And Cochran is hot on his heels, having inserted $321 million in Mississippi earmarks.[164]

Stevens has seen to it that our tax money will go for such worthy projects as $25,000,000 for rural and native villages, $1,300,000 for berry research, $1,099,000 for alternative salmon products, and $500,000 for fruit and berry crop trials.[165]

For his part, Thad Cochran's pet earmarks include $10,000,000 for the Mississippi Conservation Initiative; $5,766,000 for the Wildlife Habitat Management Institute; $1,433,000 for curriculum development at Mississippi Valley State University; $1,389,000 for the Delta Conservation Demonstration Center in Washington County, Mississippi; $936,000 for advanced spatial technologies; $517,000 for aquaculture research; $300,000 for the National Center for Natural Products; $180,000 for natural products research; and $50,000 for cotton ginning research.[166]

ACTION AGENDA

So how can the new leaders of Congress stop this earmarking insanity from gobbling up the entire federal budget? How can we stop members of the Appropriations Committee—and other powerful legislators—from spending us into oblivion?

As noted, the disclosure requirements passed by the new Democratic Congress are not the answer. Members are more than happy to disclose their earmarks—if not in the *Congressional Record,* then in campaign brochures and television ads. When it comes to earmarking in Congress, the reigning philosophy is the more the better.

We need to reform the entire process.

One brave soul, Congressman Jeff Flake, a three-term Republican Congressman from Arizona, actually rose in the House of Representatives on June 14, 2006, to ask the body to remove a dozen earmarks from the pork-laden Transportation-Treasury-HUD appropriations bill. Among his targets: $1.5 million for the William Faulkner Museum in Oxford, Mississippi; $250,000 for the Strand Theater in Plattsburgh, New York; $500,000 for swimming pool renovations in Banning, California (which happens to be the district of the Chairman of the House Appropriations Committee, California GOP Rep. Jerry Lewis); and $500,000 for a Crafton Hills College athletic facility in Yucaipa, California, also in Lewis's district.[167]

Flake's amendments were not only rejected, they were greeted with scorn. One congressman said defiantly: "We say that we know better than federal officials and bureaucrats . . . where to spend money." And Rep. Lewis, currently under criminal investigation by the Justice Department for his ties to a lobbying firm that specializes in getting earmarks passed, put Flake down saying that he "seems to have much more confidence in bureaucrats downtown than he has in the members of the House."[168]

So if Flake's approach won't work, what is the solution?

The *Wall Street Journal* has one idea: bring the president into the process.

Most recent presidents have begged Congress for the line-item veto, the ability to delete specific items in an appropriations bill. As both Ronald Reagan and Bill Clinton were fond of noting, forty state governors have this power and their liberal use of the veto is a big reason their state budgets are balanced.

HOW TO STOP EARMARKING IN TWO EASY STEPS

1. **LINE ITEM VETO:** Give the president the power to veto specific appropriations. Right now, he's able only to sign or veto an entire bill, not to redline specific elements. The government needs the money, so he's usually obliged to sign appropriations bills as they're presented to him. A line-item veto would allow him to X out the pork.

2. **LET THE PRESIDENT IMPOUND APPROPRIATIONS:** When Congress votes to spend our money on nonsense, let the president refuse to write the checks. The chief executive used to possess this power; it's time to bring it back.

The "Contract with America"—formulated by House Speaker Newt Gingrich—animated the GOP seizure of power in Congress in 1994, calling for passage of the line-item veto. Clinton alertly hopped on the bandwagon and said he agreed with this aspect of the Contract. So Congress finally passed it, only to have the Supreme Court strike it down in 1997 as violating the separation of powers between the two branches. But some advocates believe it may be possible to pass a line-item veto that would pass muster with the Court. If so, they should enact it promptly.

We also should repeal the 1974 Budget Act, which stopped the president from "impounding" funds that have been appropriated by Congress. Passed at the height of the battle between Nixon and the Democrats in Congress—a fight that ultimately contributed to the president's forced resignation—it stops the chief executive from using the power that the *Wall Street Journal* points out "every President from Jefferson to Nixon [had] to refuse to spend money if the funds were unnecessary." [169] Congress passed this restriction because it was angry at Nixon for impounding funds it had voted, and enacted the bill as a way of cutting this unpopular president down to size.

But the *Journal* is right: the president should have this power.

Will he use it well? It's reasonably certain that the chief executive would do a better job than Congress at restricting pork. After all, the White House, by definition, resisted pressure for these very projects by refusing to put them into the budget it sent to Congress—hence the need for the earmarks in the first place! It's not that the lobbyists didn't ask the Secretary of Transportation or the White House to put money in their budget requests for their pet pork. They probably lined up outside the door and demanded their pounds of flesh. But the executive branch, elected by the whole country, told them to get lost, so they went to their favorite local congressman instead and got the money.

In any event, impoundment only allows the president to veto spending. He can't add to the budget unilaterally. So since impoundment can only cut federal spending, we say, Go to it!

Earmarking is an outrage, and it must be stopped. The waste of tax money is scandalous. The legalized bribery that gets the earmarks passed in the first place is sleazy. And the projects often get in the way of important work the government must do.

If Congress doesn't act to curb this abuse, we can do it on our own. We can punish—instead of rewarding—our senators and congressmen for their pork spending. If voters stopped voting for members of Congress who bring home the bacon and understand that this practice has to stop, earmarks will vanish on their own.

So rather than just lament the sorry state of the appropriations process, let's zero in on the worst offenders in the hopes that their constituents get so fed up with their pork that they don't dare continue to rip us off.

After all, for the most part these senators and congressmen do not pass these ridiculous entitlements for personal gain. They surely get campaign donations in return for their earmarks, but not money they can put in their pockets. Their motive is *to get reelected,* either by furthering projects that give them bragging rights within their district or by getting campaign funds to run expensive television advertisements—or both.

We need to grow up and realize that the earmarks that put money into our neighborhood's right pocket force the federal government to reach deeper into our left to fund them. We have to think as national taxpayers, not just as local constituents. Once we realize the shell game the Congress is playing, we're sure to reject them for trying to fool us.

It's all a bit like inflation. During the days of double-digit price hikes, our political leaders and labor unions loved to tell us how much they were increasing our wages, pensions, and salaries. But it didn't take long before we caught on and realized that their bounty was causing the very inflation they claimed to be trying to curb. If they'd only stop trying to catch inflation, we saw, they could eliminate it—and they did.

We're like the donkey chasing a carrot on a stick, and we have to stop. It's time to realize that earmarks do not benefit us at all, but rob us of our tax money.

As voters, all we have to do is just say no. On second thought, make that *Hell, no!*

HOW ABOUT PAYING MEMBERS OF CONGRESS MINIMUM WAGE FOR THEIR MINIMUM WORK?

Congressmen and senators generally take awfully good care of themselves—especially when it comes to providing their own salaries and benefits. They're not so generous when it comes to everyone else, though. It's been nine years since Congress voted for any increase at all in the federal minimum wage, yet during that same time period congressmen's and senators' own salaries went up by about $31,000[170]—a 23 percent increase. They're now paid $165,200 per year; next year it will leap to $168,500. And, at the same time, they're doing less and less work to justify their higher salaries.

Now the House of Representatives and the Senate have passed legislation raising the minimum wage to $7.25 an hour. The Senate has insisted on coupling the increase with tax breaks for small businesses that the House may not accept. But it's likely that some compromise will finally allow an increase in the minimum wage, which has not gone up in ten years.

While congressmen carefully and regularly tend to their own economic interests, they deliberately ignore the needs of the poorest members of the workforce. In order for the lowest paid workers in our country to get any kind of raise in the woefully inadequate $5.15 federal hourly minimum wage, Congress has to take action. The minimum wage has lagged so far behind the rate of inflation that, expressed in 1996 dollars, the wage has actually dropped from $5.15 in 1996 to the equivalent of $4.04 today.

For a decade, however, Congress has refused to do anything about it. So unless these workers live in one of the states that has enacted its own higher minimum wage, their annual salary for full-time work amounts to about $10,700 a year.[171] (For a current list of state minimum wages, see http://www.infoplease.com/ipa/A0930886.html.)

At the same time, the members of Congress have made it extremely easy to make sure that they themselves get paid more and more each year, and to distance themselves from any responsibility for their self-serving actions. Neither the House nor the Senate has to do anything to ensure that they get their own annual pay increases; they're automatic. In setting up these guaranteed yearly raises, Congress was very clever. To shield themselves from likely voter wrath, Congress instituted an automatic cost-of-living increase that doesn't require a vote—and therefore doesn't leave a record. Unless each House specifically votes to decline the automatic increase, it goes into effect. This year, the increase will mean that members will receive an extra $3,300.[172] And the members can all honestly claim they never voted to raise their own pay. There's no accountability whatsoever.

Some Democrats, including Hillary Clinton, have proposed that congressional pay raises be frozen until there is an increase in the minimum wage. And she wants any future congressional salary increases to be accompanied by an equal percentage raise in the minimum wage. When Hillary was in the White House, however, Dick proposed that the president suggest that the minimum wage be indexed to the cost of living, just as congressional salaries are. President Clinton liked the idea and so did Trent Lott, the Republican Senate majority leader. But Hillary opposed it—and, after some browbeating, Bill did too. Their stated reason was that they wanted the wage to go up faster than inflation. But don't you think they really just wanted to preserve the annual fight over minimum wages and the campaign fodder it gave the Democrats?

Of course, Hillary can well afford to hold up on increases in her senatorial pay. If she got the expected cost-of-living increase next year, it would come to $3,300-not much when you consider that the Clinton family income was $8.5 million last year. Nor would it matter much to the dozens of other millionaires in the Senate.

The House bill that raises the minimum wage, by the way, provides for only a one-shot increase and no future indexing. Apparently it's all right to avoid having to legislate each increment in congressional pay, but not to avoid doing so with the minimum wage. Democrats don't want minimum wage indexing, because they want to be able to use the issue as a political football. Republicans don't want it because they can't find it in their hearts to raise it at all.

But the Republicans were even more disingenuous in their minimum wage proposals. They said they'd be more than willing to raise the wage if Congress repealed the estate tax on anyone with a net worth of up to $5 million. They were literally unwilling to give a raise to a worker making $5 an hour unless they got tax relief for those worth $5 million. Nothing could better express why people hate Republicans!

Their excuse was to say that the estate tax relief would help small businessmen pay for the higher minimum wage. But of course that is nonsense. Small businesses found it possible to pay $5.15 in 1996 when the wage was last raised; why should they object to an increase to $7, which the Democrats were proposing, when most of that increase would simply keep pace with the inflation that has eaten up almost a quarter of the 1996 wage level since it was passed?

Still, Hillary had a good idea in proposing a linkage between congressional pay raises and increases in the minimum wage. But we've got a better one.

How about paying members of Congress the minimum wage—the one that they feel is so adequate—for the actual hours they work on legislation and constituent services? (This is, after all, what they were elected to do!) Let them do fund-raising, campaigning, traveling on government and private funds, and socializing with lobbyists on their own time. Given the work schedule that they've been keeping this year, we'd see most of them looking for a new job. Pronto. As noted, the House of Representatives was only in session for 103 days in 2006. That's only two days per week. Presuming eight-hour days, they'd be paid $4,243 for the year, about 2.5 percent of their current salaries—just about the same amount as their secret cost of living increase! How many of them would be interested in that job?

Originally, members of Congress were actually paid only for the days they actually worked. From 1789 to 1815, they were paid $6 a day.[173] (See chart on page 161.) For the next two years, their salary was $1,500 per year. But, in 1817, they returned to a per diem rate of $8 per day, which remained in effect for almost forty years. Since 1855, they've been paid an annual salary—which began at $3,000 per year and has since grown to the current rate of $165,200. Since 1987, members have more than doubled their salary. And there's no reason to think these annual increases will ever stop.

At least one of the Founding Fathers expressed concern that a paid legislature would lead to ever-increasing taxes and an obsession with power. At the Constitutional Convention in 1787, Benjamin Franklin objected to a salaried Congress and warned that even a nominal payment to the members of the new body would result in higher and higher payments. Speaking to the delegates, Franklin accurately predicted the spiraling increases in congressional salaries:

> [T]ho we may set out in the beginning with moderate salaries, we shall find that such will not be of long continuance. Reasons will never be wanting for proposed augmentations; and there will always be a party for giving more to the rulers.[174]

Was Franklin right or what? Look at how congressional salaries have gone up, and compare them with the miserly increases in the minimum wage.

SENATE SALARIES SINCE 1789[175]

(all figures given are per annum unless otherwise noted)

1789–1815: $6.00 per diem

1815–1817: $1,500

1817–1855: $8.00 per diem

1855–1865: $3,000

1865–1871: $5,000

1871–1873: $7,500

1873–1907: $5,000

1907–1925: $7,500

1925–1932: $10,000

1932–1933: $9,000

1933–1935: $8,500

1935–1947: $10,000

1947–1955: $12,500

1955–1965: $22,500

1965–1969: $30,000

1969–1975: $42,500

1975–1977: $44,600

1977–1978: $57,500

1979–1983: $60,662.50

1983: $69,800

1984: $72,600

1985–1986: $75,100

1987 (1/1–2/3): $77,400

1987–1990 (2/4–1/31): $89,500

1990 (2/1–12/31): $98,400

1991 (1/1–8/14): $101,900

1991 (8/15–12/31): $125,100

1992: $129,500

1993: $133,600

1994: $133,600

1995: $133,600

1996: $133,600

1997: $133,600

1998: $136,700

1999: $136,700

2000: $141,300

2001: $145,100

2002: $150,000

2003: $154,700

2004: $158,100

2005: $162,100

2006: $165,200

Note: Since the early 1980s, Senate leaders—majority and minority leaders, and the president pro tempore—have received higher salaries than other members. Currently, leaders earn $180,100 per year.

Source: U.S. Senate

THANKS FOR THE PATRIOT ACT AND NSA WIRETAPPING

Liberals like to say that good fortune, a lag time in al Qaeda planning, normal investigative techniques in vogue before 9/11, and moderation by the terrorists are responsible for the absence of terror attacks in the United States since 9/11.

They're dead wrong.

The fact is that the record is filled with terror plots that the Department of Homeland Security and the FBI disrupted because they used the new powers and resources we gave them in the aftermath of 9/11.

Imagine what would have happened if we hadn't acted!

IF THE LIBERALS HAD HAD THEIR WAY . . .

- Terrorists would have cut the cables holding the Brooklyn Bridge aloft and the entire span would have plunged into the East River, killing tens of thousands.

- They would have attacked the New York City subway system, just as they did the London Metro; they would have blown up the Herald Square stop—where Macy's and the heart of New York's shopping area is located.

- They would have destroyed Chicago's Sears Tower, as attractive a target as the World Trade Center.

- They would have obliterated the United States embassy in Paris, at the heart of the beautiful Place de la Concorde—right next to the spot where the guillotines reigned during the French Revolution.

- They would have blown up an Ohio shopping mall near Columbus.

- They would have destroyed the Toronto stock exchange and a number of police headquarters buildings there.

- They would have succeeded in mounting a strike called Operation Crevice, leveling important buildings throughout the city of London, England.

- They would have blown up New York City's tunnels.

- A Chicago terrorist would have exploded a dirty bomb, spreading radioactivity over a large area of the United States.

- They would have attacked cargo ships taking oil from the Straits of Hormuz, through which 40 percent of the world's oil travels.

None of these plots was executed; many of them—including the final three listed here—were thwarted just in time. United States, British, Canadian, and French security services naturally don't spell out the details of their investigative techniques; we glimpse them only through a glass darkly. But sifting through published evidence in light of the provisions of the Patriot Act and of the National Security Administration's recently revealed wiretap policy, we can trace much of this progress directly to these two programs.

Unfortunately, in January 2007 the Bush administration gave in to liberal pressure and agreed to seek orders from the court set up by the Foreign Intelligence Surveillance Act (FISA). It reversed its earlier assertion that it could not get orders from the FISA court quickly enough and needed to act without warrants to protect national security. Many of us had predicted that the Left would kill the NSA warrantless wiretapping program after the Democrats took control of Congress, but few realized that the administration would concede without a fight, after four years of vigorously defending the necessity of bypassing the FISA court.

The details of the plots listed above should be enough to make any reader wonder at the folly of the administration's concession, and hope and pray that it does not increase our vulnerability to terrorist attacks—as this same administration has been telling us it would for four years.

Even as we debate the question, those who are actually on the front lines protecting us from terrorists meet little but abuse, derision, and suspicion from America's liberals and its media acolytes. But the details of how these terror plots were broken up show that what Churchill said of the Royal Air Force after the Battle of Britain in 1940 can equally be applied to the Department of Homeland Security and its counterparts in other nations: "Never . . . was so much owed by so many to so few." [1]

Brooklyn Bridge

What a tempting target for terrorists! Each day, 144,000 cars cross the Brooklyn Bridge, one of the oldest suspension structures in the world. Connecting downtown Brooklyn to downtown New York, the bridge was completed in 1883, when Brooklyn and New York were separate cities (they joined in 1899). [2]

The plot to blow up the bridge is no mere speculation: it was a genuine plan, foiled in 2002. Lyman Faris, a thirty-four-year-old Ohio truck driver and naturalized U.S. citizen who immigrated from Kashmir, was planning to use a gas blowtorch to burn through the suspension cables that held up the bridge. Faris's operation was masterminded by Khalid Shaikh Mohammed, the man who is blamed for planning the 9/11 attacks. [3] Faris had been to Pakistan and Afghanistan, had met with Osama bin Laden, and, according to CBS News, had "scouted U.S. sites for possible terrorist attacks." [4] At first, Faris met with bin Laden to provide "information on ultralight aircraft" and to help al Qaeda "obtain airline tickets and cell phones. [5]

"At a meeting after the September 11 attacks, an al Qaeda operative asked Faris about his job as a truck driver. . . . Faris told him he made deliveries to airports, including some directly to cargo planes."[6] Eventually, Faris was assigned to target the Brooklyn Bridge. He was arrested before he could make good on his attempt.

How did we find him?

The FBI actually first interviewed Faris shortly after 9/11. Remember how they fanned out across the country to interview Muslims—and how vehemently left wing groups in the United States objected? Faris was one of those targeted by those federal agencies.

According to the *New York Times,* "Shortly after the Sept. 11, 2001, attacks on the World Trade Center and the Pentagon, FBI agents in Ohio interviewed Mr. Faris and his former wife, Geneva Bowling."[7]

Federal officials said they were "drawn to Mr. Faris by several flags in his profile, including his age, his Pakistani lineage, and his work as a commercial truck driver with a hazardous-material license."[8] A law enforcement official told *The Times* that "one issue of interest [was] the fact that he had a commercial driver's license and, possibly, a hazmat attachment to that."[9]

But the trail ended there and federal officials took no further notice of Mr. Faris . . . until they picked up the scent again in the spring of 2002.

What renewed their interest in Faris?

The Feds aren't saying, but some very good evidence indicates that it may have been the very wiretapping by the National Security Administration (NSA) that had liberals up in arms.

Newsweek has reported that "a transcript of Faris's October 2003 sentencing hearing . . . makes cryptic references to . . . sources of intelligence that helped the federal investigation." Prosecutor Neil Hammerstrom Jr. told the judge that on March 19, 2003, two FBI agents and an antiterror task force officer went to interview Faris in Ohio following what the prosecutor described as "a call that was intercepted in another investigation." Hammerstrom said he didn't want "to get into too many details in open court." He made this statement at a time when the NSA wiretaps were still a closely guarded secret, before the *New York Times* decided to spill the beans and compromise it.[10]

The FBI arrested Faris, but while he was in custody he was allowed to make cell phone calls; according to his lawyers, the Feds monitored his calls—this time with a warrant—which may have added to their knowledge of his planned crime.

As *Newsweek* reported, prosecutor Hammerstrom says that the FBI got "overseas source information" that Faris had been assigned by al Qaeda to evaluate the Brooklyn Bridge as a possible target. Faris's current lawyer, David Smith, "is convinced this information came from Khalid Shaikh Mohammed, who had just been arrested in Pakistan." [11]

Did the NSA wiretaps help lead authorities back to Faris? According to the *New York Times*, officials had claimed that the warrantless wiretaps had helped to expose the plot to blow up the bridge, citing it as one of two terror operations that the taps had prevented. [12]

New York City police commissioner Ray Kelly says that he first learned the Brooklyn Bridge might be a target when federal authorities warned him that the words "the bridge in the Godzilla movie" (i.e., the Brooklyn Bridge) had showed up again and again in conversations they had been monitoring. While these officials did not reveal the existence of the then secret NSA warrantless wiretaps, it is not a bad conjecture that this was where the warning came from. [13]

How was the NSA supposed to get a warrant to find out if people were going to blow up the Brooklyn Bridge when they had no idea of the target or the plotters until *after* their monitoring of overseas calls from the United States had turned up unusually frequent mentions of the Brooklyn Bridge? Once they knew of the plot, they could—and did—get warrants to tap phones and investigate. But in the initial effort to discern patterns which might suggest future targets, it's impossible to describe what one is looking for, much less get a warrant to find it.

Once Kelly learned that the terrorists were referring to the Brooklyn Bridge, he flooded it with New York City police officers. Kelly's staff told us that federal authorities told them they had picked up a mention that the bridge was "too hot" in their intercepts, indicating that the prophylactic efforts had been successful. [14]

It's likely that 9/11 mastermind Khalid Shaikh Mohammed, then (and now) in federal custody, confirmed the plot, giving police the information they needed to raid Faris's apartment and arrest him.

So how, exactly, did the Patriot Act help to save the bridge? Because it gave federal authorities the jurisdiction they needed to interrogate Mohammed without giving him access to an attorney. It also mandated that federal authorities share their information and leads with local officials—

hence the call to Commissioner Kelly. In the past, federal agencies have often been loath to share information with local police, for fear that their sources would be compromised—or that word of the wiretapping would get out. But, because of the Patriot Act, so despised by the Left, the Feds shared the information in time and averted the deaths of tens of thousands of innocent commuters.

(How ironic that Senator Hillary Clinton, allegedly representing New York State in the Senate, voted against extending the Patriot Act in December 2005. She succeeded in blocking its renewal, but agreed to a short-term renewal pending further negotiations. After she had watered down the Act, she voted to extend it, in weakened form, in January 2006.)

Although he later tried to recant, Faris confessed and was sentenced to twenty years in prison. He is currently serving this sentence at a federal prison in Florence, Colorado.[15]

The Faris investigation also helped to expose another of his plots—this time with coconspirator Nuradin Abdi, a thirty-two-year-old Somali immigrant—to blow up a shopping mall in the Columbus, Ohio, area where they lived. Attorney General John Ashcroft alluded to the plot, saying that "the American heartland was targeted for death and destruction by an al Qaeda cell which allegedly included a Somali immigrant who will now face justice."[16]

Operation Crevice

When the *New York Times* decided to compromise national security by publishing the details of the NSA wiretap program, federal authorities told the newspaper that the wiretaps-without-warrants had been instrumental in thwarting another terror plot: a conspiracy the British called Operation Crevice.

Early in 2004, according to *Newsweek,* British authorities arrested a group of Pakistanis living in the United Kingdom "on charges of plotting to mount attacks in Britain using explosives made out of fertilizer." UK police "never established precisely what targets the plotters were allegedly going to attack." Apparently the key to cracking the case was the arrest of a former New York resident, Mohammed Junaid Babar. As *Newsweek* reported,

"Babar, a computer programmer, had left the United States after 9/11 and a few months later gave a widely broadcast TV interview, speaking in an American accent about his pro-jihad and anti-American views." [17]

The magazine said, "It is unclear how U.S. authorities knew Babar was returning to the United States in the winter of 2004—perhaps NSA monitoring played a role in tipping off U.S. authorities." [18] After his capture, he pled guilty and agreed to become a government witness. "A lengthy statement Babar gave to investigators became a key piece of evidence in the UK prosecution of Operation Crevice suspects, and in the investigation by Canadian authorities of another suspect who was working as a contract computer programmer at the Canadian Foreign ministry in Ottawa, according to three sources familiar with the case." [19]

The Patriot Act was also helpful in exposing Operation Crevice. *Newsweek* reporters learned from Justice Department sources that one of the ways they uncovered the plot was "by using controversial provisions in the Patriot Act that gave the Feds new powers to obtain records of public-library patrons. The officials indicated that investigators had monitored Babar's Internet use at a New York Public Library during which he allegedly exchanged messages with terror suspects abroad." [20]

That particular provision of the Patriot Act is among its most controversial. Privacy advocates have visions of federal investigators snooping about to find their reading habits. Weakening the federal ability to access these records was one of the amendments on which Senator Hillary Clinton insisted before she and the other Democrats would vote to extend the Patriot Act.

As we contemplate how the library provisions of the Patriot Act helped to avert a massive terror attack in London, let us remember the scornful dismissal of ultraliberal senator Russ Feingold (D-WI): "[This] provision," he said, "might permit an employer to give permission to the police to monitor the e-mails of an employee who has used her computer at work to shop for Christmas gifts. Or someone who uses a computer at a library or at school and happens to go to a gambling or pornography site in violation of the Internet use policies of the library or the university might also be subjected to government surveillance—without probable cause and without any time limit." [21]

The Missile Launchers

One of the most deadly terror attacks that NSA warrantless wiretaps possibly foiled was the plot by mosque leader Yassin Aref of Albany, New York, and Kurdish refugee Mohammed Hossain to buy missile launchers.

After learning that the pair was trying to buy the launchers, the FBI trapped them in a sting operation and arrested them, in August 2003. Aref was also charged with "having documented connections to key terrorist figures in the Middle East," according to the *Albany Times-Union*.[22]

Aref's defense attorney moved to quash the evidence of "fourteen phone calls Aref allegedly made from Albany to the Damascus office of the IMK, or the Islamic Movement for Kurdistan" on the grounds that the "conversations were illegally taped by the National Security Agency in the post-9/11 fight against terror. Published reports also have linked Aref and Hossain's alleged activities to that program. The Justice Department claims the calls made to a Syrian number between 1999 and 2001 were a means of gathering intelligence for Osama bin Laden."[23]

The Dirty Bomb Plot

The Patriot Act has also come in for special scorn from the Left because it allows terrorists like Jose Padilla, an American citizen who was detained at Chicago's O'Hare Airport, to be held as an "enemy combatant." (After a lawsuit challenging his incarceration, the Justice Department voluntarily transferred Padilla's case to civilian U.S. courts.)

What did Padilla do? He wanted to explode a dirty bomb in the United States. A dirty bomb is a nuclear device which explodes after only a partial chain reaction—not enough to produce a blast, but strong enough to scatter radioactivity over a wide geographic area.

In 2001, the BBC reports, Padilla "made contact with Abu Zubaydah, a senior al Qaeda commander who is in American custody and apparently cooperating with the FBI." Al Qaeda asked Padilla to travel to Pakistan, where "he learned how to make a dirty bomb and allegedly met several other al Qaeda members."[24]

The Patriot Act permitted us to apprehend Mr. Padilla, to hold him in custody, and to interrogate him before he could pull off his attack.

U.S. anti-terrorism operatives frequently need to act quickly to smash a terror plot before it can reach fruition. Sometimes, they must make arrests before they have sufficiently concrete evidence of the target or the plans for a terror attack. But since the Patriot Act makes it a crime to provide "material support for terrorists" (Sections 201 and 805), it is possible to arrest terror plotters in the early incubation stage of their operations. The *Los Angeles Times* noted that "the material-support law [in the Patriot Act] has been a popular legal tool used by the Justice Department to cut down what it has viewed as evolving threats against the nation." [25]

The Sears Tower Plot

The Feds used the material support provision of the Patriot Act to arrest seven people for what the *Los Angeles Times* described as "a nascent plot that allegedly involved attacks against the Sears Tower in Chicago and federal buildings in South Florida." The FBI had reportedly infiltrated the terror cell with an informant posing as an al Qaeda operative. "The men are believed to have discussed with the informant the possibility of attacking the 110-floor Sears Tower—the nation's tallest building—and the FBI office in Miami." [26]

The Patriot Act allowed the Justice Department to arrest the plotters early in the game, well before anyone was at risk from their activities.

ACTION AGENDA

But do the Patriot Act and the NSA wiretaps give law enforcement agencies the full powers that they need to keep us safe?

This question acquired special urgency over the summer of 2006, when British anti-terror officers, collaborating with American and Pakistani operatives, arrested twenty-five men charged with a plot to destroy airplanes flying from London to New York City. The British smashed the plot before it could be carried out, but the question remains as to whether or not American law enforcement officials would have had the ability to do so were the plot on U.S. soil.

PBS, uncharacteristically, aired a special investigation on the difference between American and British laws governing terrorist investigations, highlighting the respects in which UK police have powers that American investigators can only dream about.

Homeland Security secretary Michael Chertoff told PBS that American anti-terrorist investigators wanted to "have the ability to be as nimble as possible with our surveillance." Toward that end, he pointed out that the British "ability to hold people for a period of time gives them a tremendous advantage" and that "there are some legal restrictions here [in the States] under the Constitution that [the British] don't have, but their nimbleness and their flexibility are important tools we want to have here, as well." [27]

Robert Leiken, director of the Immigration and National Security Programs at the Nixon Center, told PBS that the greater British nimbleness to which Chertoff alluded came about "both because of their laws and because of the politics. They're allowed to get a wiretap without a court order. They can detain a terror suspect for as long as twenty-eight days before he has to be charged." [28] In the United States, a suspect can be held for only forty-eight hours.

Jeffrey Rosen, professor of law at George Washington University and legal affairs editor at the *New Republic,* underscored the importance of the twenty-eight-day rule. "[It's] a tremendously important difference for the British because it allows them to delay arrests to allow investigations to proceed because they don't have to develop probable cause. They know that they can swoop down and detain someone without cause at any moment." [29]

As a result, British anti-terror investigations are likely to go on longer, giving the agents the time they need to learn everything they can of the extent of the terror network before arresting the suspects and breaking up the plot. In the Sears Tower bombing arrests, for example, U.S. agencies pounced on the plotters at a very early stage, rather than let the scheme unfold as the British did in foiling the London-to-New York jetliner plot.

Rosen continues: "Once you are formally arrested in America, there are more protections at trial, greater access to lawyers. You can challenge your conviction as unconstitutional under the writ of habeas corpus, which Britain very interestingly has cut back on since 9/11. This cradle of English liberties has now been severely restricted. And that, too, is being challenged in America as the administration has asserted the right to detain people as enemy combatants indefinitely." [30]

The differences between the American and British systems also apply to securing telephone taps. Tom Parker, a former British counterterrorism official who now teaches terrorism studies at Bard College in New York State, noted that "there is a difference in the way that we apply for telephone intercepts in the United Kingdom. In the intelligence sense, these are authorized by the home secretary, by a political figure, not by a judge. But it still is a fairly rigorous process. Reasonable grounds have to be provided for the application, and it is put through a series of very stringent bureaucratic checks."[31]

The British criteria of "reasonable grounds" is, of course, far less stringent than the U.S. ground rule of "probable cause."

The plot to blow up as many as ten jetliners over the Atlantic was broken up, it would appear, by British infiltration of the terrorist group. In the UK, no court order is required to infiltrate an organization suspected of terrorism. But, as Leiken points out, "In the United States, if we want to . . . infiltrate a mosque, we have to get a court order. We have to go before a judge. You have to show at least reasonable suspicion. [The standard] used to be probable cause."[32]

Another key difference between the British and American protections against terror attack is the proliferation of cameras posted in public places throughout the UK, particularly in London. Every street corner and street in London is covered by a surveillance camera. So ubiquitous is the coverage that it is used to enforce traffic laws.

London requires motorists to pay a "congestion tax" of £7—about $14—for each day they drive in the heart of the city. Failure to pay the tax triggers a £50—about $100—fine. But if you don't pay within two weeks, the fine jumps to £150, or about $300. You don't get a ticket on the street. The camera automatically picks up your license plate, and the computer sends you the fine. Estimates claim that the average motorist is photographed five times a day on the London streets.[33]

While this obnoxious surveillance is primarily used for traffic reduction, it has proven an incredibly important surveillance tool. As Tom Parker told PBS, "video records are a fantastic investigative tool, but they are typically a post-incident tool. So this is something really that helps detectives rather than intelligence officers. But it is tremendously useful. The authorities in the United Kingdom were able to track the [London metro] bombers back to their point of origin largely through camera systems. They could see how the plot unfolded, and they could see the dry runs that took place, as well. So you have a very powerful tool." [34]

In the United States, we probably would not want to sacrifice our liberties so fully as to put ourselves under such continuous surveillance. London accepted the cameras only after IRA bombings terrified the nation. If a similar outbreak were to happen here, we would likely embrace such an option.

But we don't have to go that far. There are some additional tools that we should insist that our legislators give the police to make us safer from terrorism. Why should our anti-terror investigators need a court's permission to infiltrate a group suspected of plotting violence against us? And why should we not adopt the UK's twenty-eight-day rule, which permits longer detention of terror suspects before charges are brought? Give the police time to develop a case before they have to proceed to an indictment or release the suspect.

We also need to be vigilant to be sure that a Democratic president or more liberal Congress does not take away the tools we need to fight terrorism. We need to keep the Patriot Act on the books and not give in to liberal paranoia by eliminating library surveillance or other key provisions, which, by now, have proven their ability to help in the war on terror.

We must not let the Patriot Act be scuttled as the NSA warrantless wiretaps have been!

HOW TEACHERS UNIONS BLOCK EDUCATION REFORM

There's nothing wrong with America's schools that breaking the power of the teachers unions wouldn't cure. Through work rules, opposition to merit-based pay, and their adamant refusal to reform, unions have held our schools back for decades.

Negotiating contracts that run to hundreds of pages and cover every minute detail, teachers unions block any efforts to improve management. One school superintendent put it this way: "You have Gulliver and the Lilliputians. You've got a thousand of these little ropes. None of them in and of [itself] can hold the system down, but you get enough of them in place . . . and the giant is immobilized."[1]

Governments and taxpayers have done their part, more than doubling education spending over the past fifteen years (from $188 billion in 1995 to $418 billion in 2005). Adjusted for inflation, per-pupil spending on schools has risen 27 percent since 1990.[2]

Despite this huge increase in spending, student achievement scores in English, for example, have barely budged in this period, gaining only 1 percent,[3] and math scores have risen by only 4 to 8 percent.[4]

The extra tax money has made no appreciable difference.

And why not? Because of the way the union contracts bind our schools, hand and foot.

HANDCUFFS: HOW THE NEW YORK CITY TEACHERS UNION STOPS SCHOOLS FROM WORKING

Here are some examples from New York City's school system, according to a study by the Hoover Institution:[5]

- School administrators can't ask a high school teacher to teach more than 3.75 hours a day.

- By union contract, administrators can't ask teachers "to supervise a lunchroom or study hall, help special education students on or off the bus, help college applicants prepare their transcripts, score city-wide tests, or write truant slips."

- "One New York City teacher cannot be paid more, or less, than any other teacher at the same level of seniority, regardless of the particular teacher's talents and effort or the difficulty of recruiting a teacher for a hard-to-find position such as math or science."

- "The right to hire or not to hire a teacher is limited by teachers' 'transfer rights,' which give them first choice on a place in another school."

- "The right to fire a teacher is limited by teachers' 'retention rights' and a complex and lengthy set of due process procedures. Assistant principals have similar rights."

- Principals cannot hire, fire, or promote teachers, nor may they determine their pay, work hours, assignments, requirements, or expectations.

The chancellor of the New York City school system, Joel Klein, reports that in the past four years he has been able to fire only two of eighty thousand teachers for incompetence.[6]

John Stossel of ABC News reported that "it took years to fire a teacher [in New York City] who sent sexually oriented e-mails to 'Cutie 101,' a sixteen-year-old student." Klein said the teacher wasn't teaching while his case was pending, but that "we have had to pay him [anyway], because that's what's required under the contract." After six years of litigation, the teacher was finally fired, but only after receiving $300,000 in compensation while his case was adjudicated.[7]

As Stossel reports, Klein concedes that the New York school system included "dozens of teachers who he's afraid to let near the kids, so he has them sit in what are called rubber rooms. This year he will spend $20 million to warehouse teachers in five rubber rooms. It's an alternative to firing them."[8]

Klein came up against the unions again when he learned that there were forty-four assistant principals whom no school in New York City wanted. Able to draw from a pool of assistant principals, the principals and school boards left these forty-four out in the cold. But the union contract guaranteed them jobs. So Klein had to face the prospect that they would be forced on the schools, bumping the men and women the principals had hired and replacing them with people they did not want. Saying, "I believe that leaders need the power to choose their own management teams," Klein decided instead to create forty-four unnecessary and useless jobs to accommodate the union contract.[9] The New York Times reported that "the chancellor's office said there are at least 24 assistant principal jobs open citywide, and that 206 other assistant principals were in danger of being displaced if the 44 were allowed to begin the bumping process."[10]

"If these assistant principals cannot find work by next week," Klein said, "I am required, under the . . . contract, to force them upon you and your colleagues, sometimes bumping out the assistant principals you have carefully chosen. In certain instances, a cascade of bumps could follow, tearing up the teams that you have built over months and years."[11]

To avoid this, the chancellor said that he would create the jobs—at a cost to the city of $5 million.[12]

And it's not only in New York City that teachers unions tie up efforts to improve school quality. The trend is national. In Milwaukee, for example,[13]

- a principal cannot require faculty members to submit lesson plans weekly—or even periodically;

- principals can't require teachers to meet before or after school to write goals or objectives without negotiating a new clause in the union contract covering it;

- teachers may not be asked to attend faculty meetings more than 2½ hours a month; and

- legislation to add twenty minutes to the school day was blocked by unions.

Because of teacher-hiring laws, public schools throughout the nation orchestrate an annual "dance of the lemons." As the *San Jose Mercury News* explains, this name refers to "the annual migration of a minority of veteran teachers who either were burned out or who didn't get along. They agreed to take voluntary transfers and gravitated to low-performing schools, where principals were desperate and parents less vigilant." As the newspaper notes, it is "hard to know how widespread the problem is. But in a survey by the New Teacher Project, twenty-one percent of principals [in California] reported that a majority of teachers hired through voluntary transfers were unsatisfactory." [14]

Governor Arnold Schwarzenegger recently signed legislation to curb the "dance of the lemons" in California and to give underserved schools greater latitude to turn down bad teachers nobody else wants.

The unions like to convince us that more money is the answer. But it's not. Even adjusting for inflation and for the increasing number of children attending school, we have increased education spending year after year, decade after decade, with no significant improvement in test scores. Look at the sidebar to see how per-pupil spending, adjusted for inflation, has risen.

PAYING MORE AND GETTING LESS:

NATIONAL EDUCATION SPENDING, IN CONSTANT DOLLARS PER PUPIL[15]

(includes all federal, state and local money)

Year	Total Per-Pupil Spending
1990	$7,143
1991	$7,150
1992	$7,144
1993	$7,135
1994	$7,182
1995	$7,251
1996	$7,245
1997	$7,326
1998	$7,515
1999	$7,762
2000	$7,953
2001	$8,220
2002	$8,444
2003	$8,643
2004	$8,845
2005	$9,062

Source: U.S. Department of Education, National Center for Education Statistics

And more money hasn't fixed the problem, as the following tables show. Reading scores among fourth and eighth graders have gone up only 1 percent in the past thirteen years! Math scores went up a bit more: 8 percent for fourth graders and 4 percent for eighth graders. And to think that during this same period we've given our schools increases of more than $200 billion a year! What a dismal return!

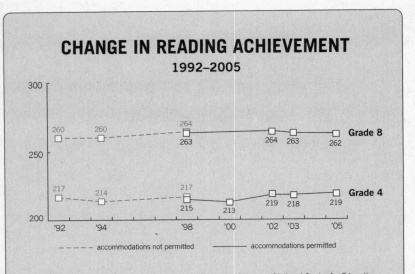

CHANGE IN READING ACHIEVEMENT
1992–2005

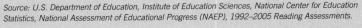

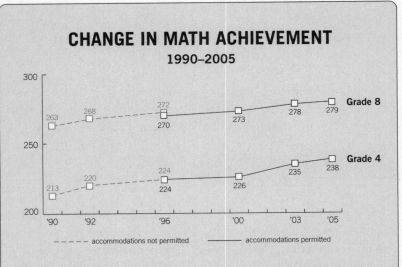

CHANGE IN MATH ACHIEVEMENT
1990–2005

And, when it comes to reducing the achievement gap between white, African American, and Hispanic American students, the public schools' record is even more pathetic. According to a review of educational testing data by Standard & Poor's, "On the whole, school districts with achievement gap reductions . . . are fairly rare."[16]

The report noted that "in the majority of the thirteen states analyzed, less than ten districts have been recognized as having significantly reduced at least one achievement gap." Yes, you read that right—ten districts out of *hundreds!* Sadly, the report concluded that "the average achievement gap in all thirteen states remains large."[17]

The bottom line is inescapable: The incremental strategies of the past few decades—spending more and more money—have done almost nothing to make our schools better.

While most Americans are conscious of the failings of our public schools, they are totally myopic when it comes to the schools in their own community, even the ones their own children attend. Schools may be bad on a national basis, they say, but not in *my* neighborhood!

The Gallup Poll asked people to rate America's public schools. No parent would like her child to bring home a report card as bad as the nation's schools got in the survey. Only 2 percent awarded America's public schools a grade of A; 24 percent gave them a B. A troubling 45 percent gave them a C, but an unrealistic 13 percent tendered a D rating, and only 4 percent gave the nation's public schools a failing grade.[18]

Even more blindly, when it came to rating their neighborhood school or the one their own children attended, most people rhapsodized about how wonderful they were! Among parents, 61 percent gave their own children's school an A or B, and only 3 percent flunked it. Those without children were almost as positive, with 47 percent awarding their local schools an A or B.[19]

AMERICA'S SCHOOLS ARE TERRIBLE . . .
BUT *OUR* SCHOOL IS GREAT!

HOW PARENTS RATE THE SCHOOLS[20]

Grade	Nation's Schools	Local School
A	3%	17%
B	19%	44%
C	44%	24%
D	13%	10%
Fail	6%	5%
Don't know	15%	0%

HOW ADULTS WITH NO CHILDREN IN SCHOOL RATE SCHOOLS[21]

Grade	Nation's Schools	Local Schools
A	2%	11%
B	26%	31%
C	45%	37%
D	13%	9%
Fail	3%	3%
Don't know	11%	9%

SOURCE: "The 36th Annual Phi Delta Kappa/Gallup Poll of the Public's Attitudes Toward the Public Schools" by Lowell C. Rose and Alec M. Gallup from http://www.pdkintl.org/kappan/k0409pol.htm.

For decades, the National Education Association (NEA), the larger of the two national teachers unions, has tried to sell two myths to the American people:

1. That our schools were in good shape and improving
2. That the only thing needed to make them even better was more money

But then President Bush proposed the No Child Left Behind Act, which Congress passed by 87 to 10 in the Senate and 381 to 41 in the House[22] (that is, with broad Democratic support). No Child Left Behind (NCLB) requires uniform student testing throughout the primary and secondary school years and mandates assessments of school and teacher performance based on these test results. Threatening to bring a results-oriented objectivity to education, this relatively new law would undercut the power of the unions. One of the best laws of the past few decades, the No Child Left Behind Act has the capacity to completely transform our education system into a high quality network of schools that meets lofty standards and raises children up to reach their full potential. From now on, schools will be rated objectively, using uniform tests all over the country. Parents will no longer have to rely on the self-congratulatory statements of school administrators, school board members, or teachers unions, but they will have objective test scores to measure student achievement.

But the teachers unions face a problem: tests administered under the act show clearly that our schools are not getting any better.

Instead of adopting reforms to improve public education, the teachers union has tried to scrap the MRI machine that diagnosed our schools' failings. Determined to discredit dismal test scores that NCLB has uncovered among our students, the union has outdone itself in a desperate, last-ditch effort to sabotage and derail the No Child Left Behind Act. Why? Precisely because it is so effective.

In the short term, these terrible test scores raise serious questions about the competence of individual teachers, highlighting those who should not be in the classroom. The union, dedicated to protecting incompetent teachers, wants no objective basis for singling out those who are cheating our kids.

But the longer term is what truly worries the union. The No Child Left Behind law requires that when a majority of students in a school score way below average, the school gets extra federal funding for tutoring, after-school services, and summer programs—to the tune of five hundred to one thousand dollars per child. But when the school *still* fails, year after year, NCLB provides that "parents with children in that school can immediately transfer their child to a better-performing public or charter school."[23]

In other words, they can do what every wealthy family in the country does: just as Bill and Hillary Clinton did with Chelsea, parents who aren't satisfied with their local public school can move their children to another institution where they can learn. Unlike the voucher system, which would allow average or low-income parents to send their children to private or parochial schools at state expense, No Child Left Behind lets parents send their children to charter schools, alternative public institutions where real learning can take place.

But charter schools are a red flag to the teachers unions. Worried that an entire new public education system may arise—without union involvement—to provide America's parents and children with a quality alternative, the unions have launched a full-scale campaign to persuade the country that charter schools don't work and don't do much to educate children.

But the data prove them wrong.

While there has been precious little improvement in public schools, charter schools are doing much, much better, largely because they tend to be free of union domination and can raise standards and pay teachers based on performance.

Most often, charter schools are sponsored by nonprofit organizations, which are responsible for one-third of the charter schools currently operating in the United States. Public schools themselves sponsor a quarter of the charters. Businesses sponsor 9 percent; parents groups, 6 percent; and other community groups account for 6 percent of the charter schools.[24]

ABC News tells the story of Ben Chavis, a former public school principal "who now runs an alternative charter school in Oakland, California." Spending thousands less per student than the surrounding public schools, Chavis "laughs at the public schools' complaints about money. That is the biggest lie in America. They waste money."[25]

Despite his lower overall spending, Chavis pays his teachers more than they would earn working in public schools. According to ABC News, "To save money, Chavis asks the students to do things like keep the grounds picked up and set up for their own lunch. For gym class, his students often just run laps around the block. All of this means there's more money left over for teaching."[26] Chavis's school "also thrives because the principal gets involved. Chavis shows up at every classroom and uses gimmicks like small cash payments for perfect attendance."[27]

Since he took over four years ago, his school has gone from being among the worst in Oakland to being the best. His middle school has the highest test scores in the city. "It's not about the money," Chavis says.[28]

Chavis issues a challenge: "Give me the poor kids, and I will outperform the wealthy kids who live in the hills. And we do it."[29]

Provoking massive national interest, charter schools have doubled in number since 2000; today, there are 1.1 million students enrolled in these institutions.[30] The schools succeed because they are innovative.

CHARTER SCHOOLS:

EDUCATION THAT WORKS

- Smaller than other public schools, charter schools provide more individual attention. The average charter school has 297 students, compared to 438 children in regular primary schools, 616 in normal middle schools, and an average enrollment of 758 in conventional high schools.[31]

- Freed from union work rules, charter schools provide more instructional hours than regular, union-dominated public schools can. A quarter have longer school days and a fifth have longer school years. Thirteen percent have both.[32]

- Charter schools spend more of their funding on instruction, typically about 80 percent. Of twenty-one full-time people on the payroll of the average charter school, seventeen were dedicated to instruction, as were 75 percent of the part-timers. In public schools, by contrast, the instructional percentage is generally in the 60 percent range. Iowa, for example, proudly boasts that it now spends 70 percent of its money on instruction, a big improvement.[33]

- Charter schools cost less. The average charter school spends $6,500 per student, whereas public schools average $8,800.[34]

Although forty states and the District of Columbia have charter schools, two-thirds of charter school students are concentrated in six states: California (216,000 students), Florida (97,000), Arizona (96,000), Michigan (87,000), Ohio (85,000), and Texas (85,000).[35]

The teachers unions, with the complicity of newspapers like the *New York Times,* have tried to besmirch the reputation of charter schools through studies that seemingly show that they make little difference. But these studies are generally among national samples, mixing states that have highly restrictive policies on charter schools with those that treat them well. Some states, under union political pressure, deliberately impair charter schools by saddling them with restrictions that doom them to fail. For example, in Connecticut, the geniuses who run the legislature allocate ten thousand dollars per pupil to public schools but only seven thousand dollars to charter schools. Other states allow only low-income or underachieving students to go to charter schools, making them a form of special education rather than an alternative educational system.

But a state-by-state analysis of comparable students shows that the charter schools are making real progress.

New York

New York City's seventy-nine operating charter schools serve twenty-two thousand students, and their records are better—often a great deal better—than public schools as a whole. In eighth grade, two-thirds of the charter schools outperformed the other schools in their district in English and also in math. In the fourth grade, three-fourths did better than other schools in the district in math, and 51 percent excelled in English. Between 2004 and 2005, public school students in the state as a whole posted only a 9 percent gain in proficiency, whereas charter school students performed 17 percent better than they had the year before.[36]

California

In California, the results are similar. Compared to their conventional school counterparts, fourth grade students attending charter schools are 8.5 percent more likely to be proficient in reading and 5 percent more likely to make the grade in math. And, when charter schools have been in operation for more than six years, their students are 12 percent more proficient in both reading and math than those in regular public schools.[37]

Washington, D.C.

Though charter schools in Washington, D.C., get one-third less funding than regular schools get, forty-seven charter schools serve thirteen thousand students in some of the poorest communities in the nation. But a recent Harvard study indicated that charter students in D.C. are 12 percent more likely than conventional school students to be proficient in reading and 13 percent more likely to reach proficiency in math.[38]

Michigan

With one of the most advanced charter school systems in the nation, Michigan's assessment tests indicate that 38 percent of students in charter schools met state math standards, compared with only 31 percent of all students. Also, 63 percent reached state reading standards, versus 60 percent of all students. Among African Americans, 46 percent of eighth graders passed math assessment tests, as opposed to only 21 percent of blacks statewide.[39]

Ultimately, the objectivity and ubiquity of the testing under the No Child Left Behind Act—and the terrible lack of progress it reveals—will force thousands of public schools to face closure because of their failure to improve, despite the larger does of federal aid earmarked for failing schools. Then, with luck, reality will dawn on school boards, parents, and even union leaders: to preserve public education, we need to rein in the teachers unions.

So far, though, the unions don't care. Resisting any effort to reward their members based on their abilities or performance, the unions are deeply committed to protecting incompetence and keeping competition, strong management, and achievement testing out of our schools. Before Congress enacted No Child Left Behind, we invariably rated educational quality based on inputs rather than results. The National Education Association (NEA) published an annual "Rankings of the States," a publication listing states in order of average teachers' salaries and total educational spending per pupil. The more a state spent, the higher it ranked in the ratings.

Bill Clinton used to campaign in Arkansas by calling on his fellow citizens to pay more in taxes to raise up the state from its forty-ninth ranking in these NEA measurements. The state's ranking gave rise to the slogan "Thank God for Mississippi," because their beleaguered neighbor to the south took fiftieth in all such ratings, sparing Arkansas the indignity of being last.

But the rankings reflected only the amount of money spent, not the educational quality this investment achieved. Imagine if we rated the effectiveness of our military based on "cost per soldier" rather than on whether they won or lost wars. Or if we assessed the excellence of our hospitals by reviewing expenditure per patient—or doctor and nurse salaries—instead of analyzing survival rates.

The system served the teachers unions well. It could catalyze a race to the top of the spending list by playing states off against one another and, for decades, succeeded in replacing measurements of performance with those based on spending.

Curiously, while the entire educational system is predicated on a grading of student performance, the NEA resisted any performance-based measurements of education. As any kid knows, the agony of quarterly grades is the key motivating factor in student achievement. But the teachers who handed out the As, Bs, Cs, Ds, and Fs staunchly resisted being graded in the same way. While every parent knew precisely how teachers rated a child's performance, no one knew a thing about the competence of the teacher who did the grading—much less the school's performance compared to other schools. Instead of relying on objective criteria, like test scores, dropout rates, or college attendance statistics in older grades, the entire system was based on a teacher's periodic, subjective judgment. But who judged the teacher? Theoretically, the principal of the school did. But the growth of teachers unions—and their protective contracts—vitiated any real power that principals may once have had.

To grasp the impotence of school principals, imagine a boss who can neither hire nor fire employees, neither raise nor lower their pay, change neither their workloads nor their schedules. Such a manager would be totally powerless. But that is the straitjacket into which the union contracts put principals. Their supervisory power is in name only.

How did the politicians ever let things get this way? Because they were terrified of teacher strikes and of the unions' voting and bargaining power. When the union demanded big pay increases, frantic mayors, governors, and school boards had to dig deep into their pockets to come up with what money they could. Finding extra money was hard. But it was easy to surrender on work rules and on limitations to principals' supervisory powers. It didn't cost the politicians anything, and they didn't need to raise taxes to pay for it. It was a whole lot easier to give in to the union than to increase school spending.

There are actually two teachers unions in the United States: the National Education Association (NEA) has 2.7 million members, and with its billion-dollar annual dues income, it rules in more school districts, with a power base in mostly rural areas and small to midsize towns.[40] Yet the American Federation of Teachers (AFT) is also powerful, with one million members, and a base mainly in the big cities—including New York, where it has practiced a militant brand of unionization for many decades. Both are powerful adversaries.

Beginning as a civic group that advocated better schools, the NEA morphed into a labor union with a more benign reputation than the AFT, which began with a series of bruising school strikes pitting legendary union leader Albert Shanker against the then mayor of New York City, John V. Lindsay.

But it was Shanker who told the truth when asked whether he considered it part of the union's responsibility to improve educational quality. "When schoolchildren start paying union dues," Shanker replied, "that's when I'll start representing the interests of schoolchildren."[41]

As unionization became pervasive, Shanker and the AFT began to become more flexible on certain aspects of educational reform. Though still deadly opposed to most proposals for vouchers or merit-based teacher pay, the AFT did begin to hint that it wanted to partner with big city school boards to upgrade education. To the surprise of many, the AFT has backed "peer review of teachers, supported the testing of teachers new to the profession, and adopted a resolution that supported some forms of alternative compensation, including higher pay for teachers in hard-to-staff schools and subjects."[42]

But the NEA was unrepentant. Its opposition to the No Child Left Behind Act can only be described as ferocious. "From the beginning, the NEA, criticizing NCLB at every turn, was blatantly opposed to the law,"[43] wrote Julia Koppich in a study of teacher unions and their approach to the NCLB. The ink Bush used to sign the new law wasn't yet dry before Reg Weaver, president of the NEA, announced that his organization would recruit a state government to act as plaintiff in a suit to block the law from taking effect. The suit would claim that NCLB violated the federal prohibition against Washington imposing costs on state and local governments without providing the money to pay for them. Weaver couldn't find a state willing to sue, but he continues to rail against NCLB as an unfunded mandate.

At the NEA's convention in the summer of 2004, Weaver called the NCLB law "a one-size-fits-all federal mandate that sets the wrong priorities—too much paperwork, bureaucracy, and testing." Weaver told the delegates: "Our schools are becoming testing factories, not centers of learning and progress."[44]

The AFT response was more constructive. President Sandra Feldman said:

> The federal NCLB Act poses yet another test of our ability to be constructive, responsive, and creative while simultaneously fighting and protecting against the indefensible. The law is built around goals we've long supported: high academic standards and achievement, eradicating achievement gaps between the haves and the have-nots, making sure that every teacher in every school is qualified, and, yes, accountability. The law also mandates reporting outcomes by student subgroup which is the right thing to do because it puts inequities out there for all to see.[45]

Whatever their legitimacy, the unions' beefs with the NCLB law are well-defined:[46]

- It relies too much on standardized testing

- It fails to take account of differences among schools

- It encourages teachers to "teach to the test"

- It doesn't do enough to recognize the special problems of urban schools

- It labels some schools as "failing" unfairly

- Its mandates on local governments are not sufficiently funded[47]

- Resources should help reduce class size, not be diverted to standardized testing

The Act "takes testing to extremes," says Reg Weaver. "All students are expected to achieve on federally required tests at the same level at the same time. It is not testing to help students. It is testing for politics. And that is wrong! Congress needs to put students first and fix the law." [48]

These objections are lame. After decades without any accountability whatsoever for our schools, and after the proven failure of the system, the NEA is quibbling about a program that finally measures student and teacher performance. Of course no one test is ideal, but it's a lot better than not testing at all!

The idea that "teaching to the test" is wrong is especially absurd when the exam measures basic reading and math abilities. We're not talking about flights of logic or philosophy classes. If teachers teach to reading and math tests for elementary and junior high school students, so much the better!

The teachers unions' real objection is not to testing or funding, but to the fact that the NCLB law may lead to a massive flight from the educational establishment to nonunion charter schools. Charter schools already educate a million students; a million more are home-schooled. And more than six million children are enrolled in costly private or parochial schools (while their parents continue to provide full support to public education through their tax dollars). So there's a very real possibility that parents will vote with their feet, moving their children out of a failing public system into successful charter schools.[49]

At last, the union has to come to grips with the fact that its own refusal to allow reform—such as merit pay, flexible work schedules, and weakened tenure—has so enervated the public school system that the entire program is in jeopardy.

At the core of this concern is the sacred cow of the educational establishment: teacher tenure. According to Koppich, tenure was originally designed to ensure that teachers were granted "due process when faced with dismissal," a step everybody agreed was "a reasonable and necessary protection for which unions fought long and hard as a way of eliminating (or at least ameliorating) the incidence of personnel decisions that were made on the basis of patronage or favoritism." But tenure has now evolved to the point where "in most states and school districts, tenure is not only easily obtained, but losing it (and therefore losing employment, even for cause) is all but impossible." [50]

Arkansas' innovative former governor Mike Huckabee has pioneered reform of teacher tenure, changing Arkansas law to make it easier to get rid of bad teachers while still protecting them against arbitrary dismissal. [51]

Teacher tenure so handicaps system managers, supervisors, and principals as to make their titles merely honorific. According to recent surveys, however, it isn't the new teachers who value tenure—it's the older teachers, perhaps conditioned by past pretenure favoritism, or perhaps losing it in the classroom and afraid of being fired—who are the ones who insist on tenure. A study by the Public Agenda found that "veteran teachers are more attached to the status quo, particularly when it comes to the kinds of job-related protections unions have provided, than are their more junior colleagues." While 57 percent of older teachers said tenure was "absolutely essential," only 30 percent of newer teachers thought so. [52]

The goal of the teachers unions and their sycophants in the Democratic Party are clear: they want to repeal the No Child Left Behind Act, or at least cripple it by emasculating the requirements for standardized testing and curbing its invitation to parents of children in failing schools to organize charter schools able to achieve excellence.

The Bush administration and the Republican Party, by contrast, want to use the dismal results of the standardized tests to show how injurious the unions' insistence on tenure and opposition to merit pay is for public education.

And the teachers union has the political clout to make its point of view stick. In the 2003–04 election cycle alone, the National Education Association and its affiliated political action committees and campaign funds spent $972,000 influencing elections. The American Federation of Teachers went one better and spent $1,871,000 on elections during the same period. [53]

Now that the Democrats have taken over Congress, will they try to reverse the NCLB Act? Probably not, because they can't. But if they capture the White House and Congress in 2008, they may well try to do so. Even then, they may not succeed. Once parents have learned how badly their schools are performing—and how poorly their sons and daughters are doing in these failing schools—the groundswell for some alternative schooling, whether through charter schools or vouchers, is likely to grow. Parents will not happily consign their kids to failing schools for long.

At the same time, the growing power of newer and younger teachers, eager for the increased compensation merit pay would offer, may undercut the iron grip of union leaders dedicated to maintaining the current inadequate status quo.

After all, the union leaders have a dirty little secret they'd prefer to keep quiet: While education spending has more than doubled in the past fifteen years, and constant dollar-per-pupil spending on public education has increased by 26 percent, teachers' salaries have remained remarkably constant!

It is a total myth—again perpetrated by the unions—that their hardball negotiating tactics have resulted in any improvement in teacher salaries. They haven't. In 1990, teachers' salaries, adjusted for inflation, averaged $47,357; in 2005, they were still only $47,750. So, as per-pupil inflation-adjusted spending rose by 26 percent, teacher pay inched upward by *less than 1 percent* over the past fifteen years!

Look at the numbers:[54]

WHAT GOOD ARE THEY?

Teachers Unions Can't Even Raise Teacher Pay

AVERAGE TEACHER PAY: GOING NOWHERE

Constant 2004–05 dollars	
1989–90	47,354
1990–91	47,357
1991–92	47,245
1992–93	47,113
1993–94	46,852
1994–95	46,742

(continued)

Constant 2004–05 dollars *(continued)*	
1995–96	46,703
1996–97	46,374
1997–98	46,636
1998–99	47,234
1999–00	47,339
2000–01	47,509
2001–02	48,043
2002–03	48,185
2003–04	48,159
2004–05	47,750

Source: National Center for Education Statistics (NCES), "Estimated average annual salary of teachers in public elementary and secondary schools: Selected years, 1959-60 through 2004–05," http://nces .ed.gov/programs/digest/d05/tables/dt05_076.asp.

Teacher pay is much too low in America. The teachers unions have not delivered. According to the U.S. Census Bureau, the median income for a four-year college graduate in the United States is $54,689, about 15 percent higher than average teacher pay (even though almost all teachers are college graduates).[55]

In fact, teachers only earn about the same as the median income for all American families, regardless of their educational attainments: $46,500 in 2005. Teachers deserve a huge increase in pay, but only in return for two new policies: pay increases should be merit-based and rigid tenure must end. A system with performance-based pay—without tenure—as we have in many of the charter schools, would help good teachers get the raises they deserve, while showing the door to bad or stagnating teachers.

After all, the current system—in which we pay all teachers inadequately while keeping those who aren't doing much good on the job—helps nobody.

ACTION AGENDA

1. Check out your school's ratings on the standardized achievement tests. If it's classified by the federal Department of Education as a

"failing school," it deserves to be mothballed! Meet with other parents and civic leaders to discuss establishing a charter school to replace the institution that is shortchanging your kids. Contact the Center for Educational Reform (www.edreform.com) and ask them to help you establish a charter school. They will put you in touch with successful charter schools in your area that can serve as models for you!

2. Under the No Child Left Behind Act, you don't have to be rich to take your child's education into your own hands—just be determined and committed!

3. If you can't opt out of the system because your school is only marginally failing your kids, you can still take matters into your own hands! Remember that, in most areas, democratically elected school boards run the schools. Field a reform slate for your local school board elections. Typically, school board contests are dominated by teachers and the unions who are well organized to mobilize their forces in these low turnout contests. But you can beat them at their own game!

4. Reach out to other parents in the district and don't be afraid to make common cause with your neighbors. Even those without children in school still pay taxes, and probably resent wasting their money on schools that don't perform. You know those folks who urge people to vote against school bond issues? Form a coalition with them. You want better schools; they want to avoid paying for wasteful institutions. Join forces, and you can defeat the union hierarchy!

THE NEW DRUG DEALERS:

How Pharmaceutical Companies Are Dangerous to Our Health

There is a new generation of drug pushers in our country. We're not talking about teenage gangsters or newly arrived illegal immigrants; no, these new pushers are our top pharmaceutical companies themselves. They don't hang out on the corner hawking cocaine, heroin, Ecstasy, or joints. They advertise and promote drugs like Vioxx, Celebrex, Bextra, and Neurontin. Because of overt advertising, sales pressure, and hidden payments to induce doctors to prescribe certain drugs, Americans are dosing themselves at a more rapid rate, and at greater cost, than ever before.

The massive push by the drug industry for ever greater sales not only leads to overmedicating the public, but it also drives up medical costs to absurd levels. When President Clinton took office, he warned that medical inflation was threatening to eat up 12 percent of our national economy. Now it has soared to 16 percent. Look at how it has shot up in recent years:

WHY YOUR INSURANCE RATES KEEP GOING UP:

Health Care Cost Increases by Year

Year	Costs as percentage of GDP
1996	13.7
1997	13.6
1998	13.6
1999	13.7
2000	13.8
2001	14.5
2002	15.3
2003	15.8
2004	15.9
2005	16.0

Source: "National Health Expenditures Aggregate—Selected Calendar Years 1960–2005," U.S. Department of Health & Human Services, available at http://www.cms.hhs.gov/NationalHealthExpendData/02_NationalHealthAccountsHistorical.asp#TopOfPage.

Remember that the United States has to compete with other countries in the world that spend vastly less of their national wealth on health care. For example, while we spend $5,274 per person on medical services—number one in the world—Germany does quite nicely spending only $2,631. France spends $2,348 and Canada spends $2,222.[1] Competing with businesses in these other countries, American companies carry a burden that slows them down and hurts our economy: health care costs.

The rise in drug spending has been astronomical. In just the past four years, American drug companies have doubled their sales. Now they sell us $250.6 billion—in a single year—in meds we may or may not need.[2] One dollar in ten spent on health care pays for prescription medicines.

Did we need to double our drug spending because we are we sicker? No. As a nation, we are healthier. Fewer of us are smoking, and we are living longer. For the past few decades, life expectancy has risen by one year every five years.

Do we need the drugs? Some, yes. But many, many of these drugs are sold just because the drug industry hustles them more than any street corner drug pusher ever did. Tens of millions of Americans are taking drugs they don't need, for conditions they won't cure. Why? Because doctors are being paid to over-prescribe medications, and because patients are getting deluged by mass television advertising urging them to take more and more drugs. Greed, combined with induced hypochondria, has hooked us on prescription medications.

Luring people to take prescription medicines, the drug industry has increased its spending on direct-to-consumer media advertising five-fold in the past ten years, from a mere $800 million in 1996[3] to $4.2 billion in 2005.

THE NEW DRUG PUSHERS

The Drug Companies . . .

- spent $800 million on advertising in 1996, $2.7 billion in 2001, and $4.2 billion in 2005,[4]

- account for 5.8 percent of all advertising dollars, and

- employ 90,000 sales representatives to push drugs to physicians, fielding one salesperson for every 4.7 doctors.[5]

Source: compiled from Christopher Rowland, "Drug Ads Deliver A Few Side Effects," Boston Globe, June 12, 2003; "Prescription Drugs: Improvements Needed in FDA's Oversight of Direct—to—Consumer Advertising," U.S. Government Accountability Office, November 16, 2006, available at http://www.gao.gov/docdblite/details.php?rptno=GAO—07-54; and Henry A. Waxman, J.D. "The Lessons of Vioxx—Drug Safety and Sales," New England Journal of Medicine, June 23, 2005, available at http:/content.nejm.org/cgi/content/full/352/25/2576?query=TOC

The FDA has only itself to blame for this avalanche of drug advertising. Citing free speech, it relaxed rules against drug advertising and opened the floodgates in 1997 when it lifted restrictions on pharmaceutical advertising.[6] Drug companies generate massive advertising revenue, and the news media—print and electronic—have a vested interest in keeping the dollars flowing into their frequently lean coffers. With almost 6 percent of all national advertising being purchased by the drug companies, they've become an important element of every newspaper's or television station's bottom line.[7]

But overt drug advertising is minimal compared to the covert gift giving and outright—though genteel—bribery of doctors to get them to do the drug industry's bidding. The nearly one hundred thousand salespeople employed by American drug manufacturers use gifts, promotional materials, outright bribes, and other incentives to convince MDs to prescribe more and more drugs, at greater and greater cost to their patients and their insurers.

The drug industry shrugs at charges that its massive campaign of national advertising for prescriptions drives people to take overly costly and unnecessary drugs, saying that consumers can't take these medications without a doctor's prescription. But then it showers doctors with inducements to get them to overprescribe medications and spends millions more to influence politicians so they won't curb their practices.

At the very least, your family doctor may be prescribing certain medicines from certain companies because of the gifts and bribes with which drug companies bombard him. At worst, he may be prescribing medication not because he feels you need it, but because he's being paid to do so.

Dr. Peter Gleason, a Maryland psychiatrist, went too far. According to the *New York Times*, he was arrested in March 2006 for "doing something that has become common among doctors: promoting a drug for purposes other than those approved by the federal government." The Feds say that "at hundreds of speeches and seminars where he was rewarded with generous fees, Dr. Gleason advised other physicians that a powerful drug for narcolepsy could be prescribed for depression and pain relief." The prosecutors charged that "in doing so he conspired with the drug's manufacturer to recommend it for potentially dangerous uses." (Dr. Gleason has admitted that he received more than $100,000 last year from Jazz Pharmaceuticals, which makes Xyrem, the drug he was hawking.[8])

Such doctors, who overprescribe or misprescribe medication, have made the American pharmaceutical industry the second most profitable business in the United States, earning 16.2 cents in profit for each dollar of sales during the period from October 2000 through September 2005. It is second only to the banking industry (which made 17.3 cents). The

Atlantic reports that "in 2002, according to Public Citizen, a nonprofit watchdog group, the combined profits of the top ten pharmaceutical companies in the Fortune 500 exceeded the combined profits of the other 490 companies."[9]

Drug companies say that they need to achieve a large sales volume so they can spend money on research to find new medications and cures for disease. But the fact is that, while drug companies make a 16.2 percent profit, they spend only 17.2 percent of their money on research.[10] It is their lust for profits, not for research, that drives their efforts to hype drug sales.

The following chart compares the sales volume and research spending of the top ten drug companies:

RAKING IT IN:

The Top Ten Drug Companies[11]
(2004 data in billions of dollars)

	Sales	Research Spending
Pfizer	50.9	7.5
GlaxoSmithKline	32.7	5.2
Sanofi-Aventis	27.1	3.9
Johnson & Johnson	24.6	5.2
Merck	23.9	4.0
Novartis	22.7	3.5
AstraZeneca	21.6	3.8
Hoffmann-La Roche	17.7	5.1
Bristol-Myers Squibb	15.2	2.5
Wyeth	14.2	2.5
Total	250.6	43.2

Percent spent on research: 17.2 percent
Percent profits: 16.2 percent

Source: Wendy Diller and Herman Saftlas, "Healthcare: Pharmaceuticals," Standard & Poor's Industry Surveys, December 22, 2005

The profits, per se, are not the problem. The difficulty is the way these companies are making the money—and the medical consequences our population faces because of their marketing techniques.

The minute a drug is approved by the Food and Drug Administration (FDA), corporate drug pushers wage a massive campaign to induce millions of people to start taking the medication. But the FDA cannot really know whether a drug is safe until it has been on the market for a while, when a large enough number of people have taken the drug, over a long enough time period, to reveal what side effects it may cause. The prerelease clinical trials on which the FDA bases its approval to market the drugs are, of necessity, conducted over a short time period with a limited statistical sample of patients.

In the past, when bad side effects emerged from new, FDA-approved prescription medications, the number of people who suffered was usually relatively small. As David Kessler, the former head of the FDA under President Clinton, has observed, "The way it used to be, if a drug got approval, its use would increase gradually over time," so that only a limited number of people were exposed to the potential side effects. With the massive marketing bombardment that now accompanies any important new medicine, however, the number of people exposed to dangerous side effects swells exponentially. Now dean of the School of Medicine at the University of California, San Francisco, Kessler notes that the extensive use of relatively untested drugs means that "many more people are going to be exposed" to potential side effects—a situation he describes as a "nightmare." [12]

The seduction of doctors to get them to prescribe specific medications ranges from the blunt and clearly corrupt to the subtle and hard to define. While the American Medical Association regulations "suggest" that doctors should not accept gifts of more than a hundred dollars in value[13] this guideline is more honored in the breech than in compliance.

Despite the guidelines, a 2001 study by the Henry Kaiser Family Foundation found that almost two-thirds of all physicians had accepted free meals, travel, or tickets to entertainment events from drug companies hoping they would promote their medications. Even more disturbing, one doctor in eight got financial or other in-kind benefits from drug reps, or received financial incentives to participate in clinical trials.[14]

Dr. Robert Cowan, medical director of the Keeler Migraine Center in Ojai, California, typifies the attitude of honest doctors when he says that he is grateful for the free medication samples that drug reps provide. But he says that their promotional practices sometimes "cross the line." And Cowan is clear about what he won't tolerate: "Asking me outright to write more prescriptions, not to use another drug, or to use the drug in an off-label fashion (for a non-FDA approved, non-studied indication), or . . . offer[ing] an outright bribe, like 'if you write a hundred scripts for my drug I might be able to get you tickets to the Stones,' crosses the line," he says. [15]

Unfortunately, many doctors don't see it that way. Frequently, the pharmaceutical companies overtly bribe doctors. According to Gardiner Harris of the *New York Times,* the drug company Schering-Plough is now under investigation by the United States Attorney's office in Boston for doing just that. The *Times* reported that at least three doctors received checks for ten thousand dollars or more in the mail, unsolicited, from the pharmaceutical firm. The checks were accompanied by a "consulting" agreement that "required nothing other than [the doctor's] commitment to prescribe the company's medicines. Some of the checks ranged to six figures." [16]

According to the newspaper, Schering-Plough also invited doctors to take part in company-sponsored clinical trials that "were little more than thinly disguised marketing efforts that required little effort on the doctors' part. Doctors who demonstrated disloyalty by testing other company's drugs, or even talking favorably about them, risked being barred from the Schering-Plough money stream." [17]

Anticipating the results of the federal investigation, Schering-Plough has set aside $500 million to pay anticipated fines. But that enormous sum makes hardly a dent in the company's $8.3 billion of sales in 2003. [18]

And Schering-Plough is far from the only drug company that engages in shady practices. Parke-Davis agreed to pay $430 million when it recently pled guilty to criminal conduct in marketing Neurontin, an anti-seizure drug. The FDA approved Neurontin to supplement treatment of epilepsy patients, but the drug "was aggressively marketed it to treat a wide array of ailments for which the drug was not approved," according to the Justice Department. [19]

Two other drug companies—AstraZeneca and TAP Pharmaceuticals—

had to pay fines totaling $1.2 billion for fraudulently inducing physicians to "bill the government for some drugs that the company gave the doctors [for] free."[20]

THE VIOXX DECEPTION

The most heinous examples of nefarious promotion of drugs occur when a pharmaceutical company has reason to suspect that its medication may be dangerous. The work of Merck to promote Vioxx, even after clinical trials suggested that it was not safe, deserves special contempt.

As the *San Francisco Chronicle* has reported, "the painkillers Vioxx, Celebrex, and Bextra may go down in history as a classic example of the danger posed by aggressive industry promotion of prescription drugs to both patients and doctors." Initially, these were touted as "breakthrough" arthritis medications that would avoid the digestive ailments other pain killers aggravate. In 2003 alone, the drug companies spent more than $1.5 billion to promote these drugs through advertising to patients and marketing to doctors.[21]

Before their dangers were proven and the drugs withdrawn from the market, twenty million Vioxx prescriptions had been written, bringing in $1.8 billion in sales. Twenty-four million Celebrex scripts were issued, generating $2.6 billion. And ten million Bextra prescriptions were issued, accounting for $900 million in sales.[22]

Democratic congressman Henry Waxman—one of the best and most courageous members of the House of Representatives—tells the story of how Merck & Company promoted Vioxx despite increasing evidence of its potentially lethal side effects.

THE DANGERS OF VIOXX ARE EXPOSED . . .

A year and a half after Vioxx was approved by the FDA, the Vioxx Gastrointestinal Outcomes Research (VIGOR) Study was published on November 23, 2000. As Waxman notes, the "study . . . showed that the patients who were given rofecoxib (the clinical name for Vioxx) had four times as many myocardial infarctions as those who were given naproxen (the alternative medication analyzed in the study)."[23]

. . . but Merck Pushes It Anyway

Surely a responsible company would halt—or at least slow down—its marketing efforts at the first sign of such trouble, no? Well, Merck did just the opposite: it actually *accelerated* its promotional activities. By the time Vioxx was withdrawn from the American market in September 2004, after another study confirmed its cardiovascular risk, Waxman reports that "more than a hundred million prescriptions had been filled in the United States . . . tens of millions . . . for persons who had a low or very low risk of [the] gastrointestinal problems" Vioxx was designed to avoid.[24]

On February 7, 2001, the Arthritis Drugs Advisory Committee of the FDA met. Reviewing the VIGOR study, the committee voted unanimously that physicians "should be made aware of VIGOR's cardiovascular results."[25]

. . . and Then Covers It Up!

Merck reacted more like a tobacco company anxious to conceal the harm its product inflicted than a pharmaceutical firm established to promote health. Waxman relates how "the next day, Merck sent a bulletin to its . . . sales force of more than three thousand representatives. The bulletin ordered: 'DO NOT INITIATE DISCUSSIONS ON THE FDA ARTHRITIS ADVISORY COMMITTEE . . . OR THE RESULTS OF THE . . . VIGOR STUDY.'" If a doctor asked about it, the sales rep was instructed to note that the study found a gastrointestinal benefit associated with taking Vioxx, and to decline to discuss the matter further.[26]

Merck prepared a pamphlet it called "The Cardiovascular Card," which "indicated that rofecoxib [the generic name for Vioxx] was associated with one-eighth the mortality from cardiovascular causes of that found with other anti-inflammatory drugs."[27] As Waxman notes, however, the card was misleading, including no data from the VIGOR study and basing its findings on studies conducted before Vioxx was approved, studies in which patients took low doses of Vioxx for short periods. The FDA noted that it had "serious concerns" with the methodology of this research.[28]

Merck made clear that any sales representative that distributed the results of "studies that raised safety questions about drugs [was committing] 'a clear violation of Company policy.'"[29]

It might as well have worded the instructions more directly: "Telling the truth about Vioxx is a clear violation of Company policy."

The drug companies maintain that they use sales representatives to educate doctors on their new medications. But their tactics resemble more a carefully choreographed seduction, one designed to preserve the illusion of innocence while actually bribing the prescribing physician to write more orders for their product.

Since 1996, the drug companies have doubled their force of sales representatives. Carl Elliott writes in the *Atlantic* that the sales reps are "so friendly, so easygoing, so much fun to flirt with that it is virtually impossible to demonize them. How can you demonize someone who brings you lunch . . . remembers your birthday, and knows the names of all your children? Doctors look forward to the arrival of the UPS truck bringing a takeout delivery for the staff, trinkets for the kids, and . . . drug samples on the house."[30] These gifts can include sports tickets, television sets for waiting rooms, or visits to tropical resorts—anything to make a sale. "One rep told me he set up a putting green in a hospital," Elliott reports.[31]

Drug companies often pay sales representatives based on the number of prescriptions their doctors write for the company's favored drugs, and the companies have invested substantial sums on information technology to track the number of prescriptions each doctor writes for their drugs. Sales representatives single out those who are particularly prolific with their pens for special treatment and favors. Elliott writes of "an arms race of pharmaceutical gift giving in which reps were forced to devise ever new ways to exert influence. If the Eli Lilly rep was bringing sandwiches . . . you brought Thai food. If GSK flew doctors to Palm Springs for a conference, you flew them to Paris."[32]

One sales representative told Eliott that "the trick is to give doctors gifts without making them feel that they are being bought. Bribes aren't considered bribes. This, my friend, is the essence of pharmaceutical gifting."[33]

Sales reps are constantly inventing new ways of delicately and diplomatically bribing doctors to prescribe their products. They offer doctors consulting contracts, paid membership on advisory boards, and salaried speaking engagements. In 2003, drug companies provided 90 percent of the $1 billion spent on continuing medical education events, mandatory seminars doctors must attend to maintain their licenses.[34]

And, in a new twist, drug companies are using these medical education events as a way to pay off doctors who do their bidding. Those who write a lot of prescriptions find themselves retained by the drug companies to present lectures to their fellow professionals on the virtues of a particular drug. It's a double whammy for the drug companies: they get a respected physician to shill for them with his colleagues, while finding a legal and dignified way to pay him back for prescribing their drug. In 2004, according to the *Atlantic*, "nearly twice as many educational events [were] led by doctors as by reps. Not long before the numbers had been roughly equal." [35]

Of course, the drug companies know what they're doing when they hire doctors to lecture other medical professionals about the virtues of their medications. When Merck hired doctors to lecture about the advantages of Vioxx, a company study found that doctors who attended these lectures were four times more likely to prescribe Vioxx than those who went to sessions where the speaker was a non-doctor sales representative.

SCAMMING THE AMERICAN MEDICAL ASSOCIATION

Lately, drug companies have been promoting so-called scientific research to push their products. Twice in the past few months, Dr. Catherine D. DeAngelis, the editor of the American Medical Association *Journal*, has said that she was misled into publishing articles about medical advances that were based on research conducted by doctors who were paid by drug companies. The doctors did not disclose their payments, maintaining that the articles were simply the product of their own independent research. Dr. DeAngelis said she would not have published either article had she known that the doctors were being paid as consultants to the drug companies. [36]

But getting individual doctors to prescribe certain meds is only one part of the industry effort to promote their products. Drug companies also spend massive amounts to run advertisements aimed at patients to get them to ask their doctor for certain drugs. A Harvard-MIT study attributed 12 percent of the growth in national prescription drug spending to paid advertising and marketing. For every dollar spent on direct advertising to consumers, the study estimated, the drug companies were able to sell $4.20 of drugs—a ratio any corner drug dealer would envy. [37]

The avalanche of drug advertising began when the FDA, citing free speech, relaxed rules against drug advertising. Anxious to hype their sales, the drug companies rushed in through the open door. Since then, television advertising of drugs has grown rapidly with no control or policing.

Recently, the drug industry—acting through its front organization, the Pharmaceutical Research and Manufacturers of America (PhRMA)—has adopted a "voluntary code of conduct for the advertising of prescription medicines on television and in print. The *New York Times* noted that "one purpose" for the code "is to fend off more stringent federal regulation."[38] As a pharmaceutical executive told *Advertising Age,* "Better to self-regulate than to have someone else tell you how to conduct your business."[39]

Since tax money generates 39 percent of all health care dollars in the United States—through Medicare and Medicaid—drug companies have been particularly active in stopping state and federal regulators from forcing lower drug prices or cracking down on over-prescribing. To make the politicians pay attention, the pharmaceutical industry is one of the biggest spenders and contributors in American politics.[40]

During the past four election cycles, the industry has given nearly $150 million in contributions to federal and state candidates and parties, along with some political nonprofits, according to a Center for Public Integrity analysis of campaign finance records.[41]

While two-thirds of the money went to Republicans, Democratic elected officials and candidates with political clout were also treated well. See the sidebar for a list of the top twenty benefactors from the drug company's largesse.

PROTECTING THE DRUG PUSHERS:

Top Twenty Recipients
of Drug Company Donations[42] (1998–2004)

President George W. Bush	$798,732
Rep. Mike Ferguson (R-NJ)	$457,967
Sen. Arlen Specter (R-PA)	$376,549
Sen. Orrin Hatch (R-UT)	$351,424
Sen. Richard Burr (R-NC)	$344,402
Sen. John Kerry (D-MA)	$305,203

Rep. Nancy Johnson (R-CT)	$301,790
Rep. Dennis Hastert (Speaker, R-IN)	$275,050
Rep. Bob Franks (R-NJ)	$269,316
Rep. Bill Thomas (R-CA)	$240,450
Rep. Rick Lazio (R-NY)*	$224,138
Rep. John Dingell (D-MI)	$214,500
Sen. Robert Torricelli (D-NJ)	$214,253
Sen. Joseph Lieberman (D-CT)	$203,850
Sen. Chris Dodd (D-CT)	$198,725
Rep. Rodney Frelinghuysen (R-NJ)	$187,453
Rep. Billy Tauzin (R-LA)**	$184,593
Sen. Rick Santorum (R-PA)	$179,900
Rep. Roy Blunt (R-MO)	$174,522
Rep. Joe Barton (R-TX)	$170,306

* Lazio was Hillary Clinton's 2000 opponent for the Senate from New York State
** Tauzin went on to become CEO of PhRMA, the drug company lobbying organization

Source: "Checkbook Politics: Over the last seven years, the pharmaceutical industry has given $150 million in campaign contributions," Center for Public Integrity, July 7, 2005, at http://www.publicintegrity.org/rx/report.aspx?aid=720 with data from the Center for Responsive Politics at opensecrets.org.

In addition, since 2000 PhRMA has given more than $10 million to so-called 527 organizations that spend money to elect candidates at the federal level.[43]

What did their money buy?

SCAMMING MEDICARE

When the Bush administration proposed extending Medicare to offer prescription drugs, waves of joy and fear alternatively swept through the pharmaceutical industry. Government payment for drugs would mean huge new sales and larger profits—but also the potential for government controls over costs. They worried that the government might negotiate lower prices with the drug industry, using the immense buying power of the federal government to accomplish a de facto price control.

But the absolute worst of all possible worlds—for the drug companies—would be for the government to demand that Medicare doctors prescribe generic drugs instead of the high-priced brands pushed by pharmaceutical companies. Generics, fully as effective as the brand drugs, would sharply reduce the cost, cut company profits, and might even create a precedent for state-run Medicaid programs looking to save money as well.

PhRMA went to work. Patrolling the halls of Congress and the White House, they got the Bush administration to agree to channel the drug benefit through private insurance companies, which could offer any plans they wanted. There would be no government requirement to use generic drugs or curb prescriptions. And the legislation, incredibly, included a specific prohibition stopping the government from negotiating with drug companies for lower prices. The pharmaceutical firms knew that, with the government on the sidelines, they could play insurance companies off against one another to keep prices high. The drug industry was able to assure that the Medicare drug benefit would line their pockets without subjecting them to cost controls or other inconveniences.

When the Democrats took over Congress in January 2007, they vowed to reverse this outrageous prohibition against the government negotiating with drug companies to lower the price. The Bush administration claimed that private insurance plans were better able to negotiate than the federal government. But Democrats answered, reasonably, that it would be better if both levels negotiated. So, on January 12, 2007, the Democratic House passed legislation requiring that the Feds negotiate with Medicare drug providers.

Republican Whip Roy Blunt (R-MO) said that the legislation meant that "a government bureaucrat will be empowered to determine what kinds of drugs our seniors have access to."[44] But the legislation does not permit the federal government to ban any drug from the Medicare system, only to negotiate a lower price. Nevertheless, Bush press secretary Tony Snow said, "If this bill is presented to the president, he will veto it." That's very unfortunate.[45]

But an Oakland, California, nonprofit organization, Health Access California, was not so easily deterred from attempting to cut the cost of drugs on the state Medicaid program. It proposed two ballot initiatives that would force state government to rein in drug spending in the state's Medicaid program.

By gathering petition signatures, the civic group put Propositions 78 and 79 on the 2005 ballot. Proposition 78 directed the state to negotiate drug discounts for low—and moderate—income families. Even worse for the drug industry, Proposition 79 ordered the state to shut out of the Medicaid program any firm that refused to provide its best price to Medicaid patients. Simply stated, this meant that pharmaceutical companies would either have to offer generic drugs to Medicaid patients, or lose out on a $3 billion a year market.

The pharmaceutical industry went nuts, raising $10 million to defeat the propositions. PhRMA vice president Jan Faiks said, "We take [the proposed initiatives] as such a serious threat to the health and welfare of the pharmaceutical industry that we have to make a stand here," in California. "It's a very bad precedent. You're the leader in the country, and there are twenty-six states that allow [such] ballot initiatives."[46]

PhRMA eventually persuaded California's governor, Arnold Schwarzenegger, to develop an alternative "voluntary discount plan called California Rx." They hired the state's top Democratic and Republican strategists to fight the propositions, and even hired former State Assembly speaker Willie Brown—supposedly a progressive Democrat—as their lobbyist.[47]

THE DRUG COMPANIES, THE LAWYERS, AND THE UNIONS CUT A DEAL

The two California ballot propositions would have slashed state drug spending, and the PhRMA lobbyists were desperate to stop them from passing. So they cut a three-way deal with trial lawyers and the public employee unions to kill the bill and to work together to bilk the public.

When the California Trial Lawyers were considering backing the propositions, PhRMA threatened to throw its weight behind a proposition limiting lawyers' legal fees. And when public employee unions seemed about to sign on, PhRMA said it would retaliate by backing a measure to require the unions to ask their members before spending dues on political activities.

But as long as these two groups played ball with PhRMA and let the drug companies rake in huge profits by milking the taxpayers, the lawyers could continue to charge high fees and the unions could spend their members' dues as they wished. They sealed the corrupt three-way bargain!

The drug industry's hard and ruthless opposition to these ballot propositions, each designed only to save the state taxpayers' money, proved highly successful. Proposition 78 lost by 1.3 million votes; Proposition 79 fared even worse, finally defeated by 1.6 million.[48]

When people ask why drug prices are going up then, or why taxes are so high, the right answer is that people are too ill-informed to vote their real interest.

But the drug companies are also waging their crusade against generic drugs in the marketplace and through the courts. The *New York Times* dates the emergence of generic drugs to 1984, when Congress passed the Hatch-Waxman Act, a compromise that reflected Congress' desire "to infuse low-priced generic competition into the high-priced prescription drug business, yet not wanting to strip Big Pharma of its incentive to bring new drugs to market."[49]

Under Hatch-Waxman, patent-protected drugs would still have "monopoly status, which meant high prices, just as they always had. But after a certain amount of time passed, generic companies were permitted, indeed encouraged, to file an application with the agency [to produce] a generic version while also mounting a legal challenge to the drug's underlying patents." A successful lawsuit meant that the generic could market the drug, without other firms offering their own generic competitors, for six months. But "even if [the lawsuit] failed to break the patent . . . if the company was first in line with its drug application and legal challenge, it still got the right to market the drug exclusively for six months, though it had to wait for the patent to expire." Drug patents usually last twenty years; as the *Times* reports, "by the time a drug gets to market, the patents usually have twelve to fourteen years remaining."[50]

The *Washington Post* has reported that "53 percent of prescriptions are now filled with generic drugs, [but] they account for only about 12 percent of [the money spent on] drug purchases"[51]And the number of generic drugs is bound to increase, as health insurers and the companies, unions, and individuals who pay their premiums cast about for cheaper alternatives.

THE SOLUTION TO HIGH DRUG PRICES:

Generic Drugs

- 53 percent of prescriptions, but . . .
- only 12 percent of the cost

Source: Washington Post

The battle between big drug companies (who want to keep prices high) and generic pharmaceutical companies (who want to exploit the market for lower price substitutes) is only now really heating up. The *New York Times* notes that the 1980s and '90s were "an extraordinary time in the drug industry, with one blockbuster drug after another brought to market."[52] Now the twenty-year patents for these drugs are expiring, and the drug companies are redoubling their battle to keep the generics off the market to keep prices high—and they have a variety of nefarious strategies for doing so.

Starting in the 1990s, pharmaceutical companies began to negotiate deals with generic drug makers to keep them off the market. In return for large sums of money, the generic makers agreed to keep their medicines off the market and not compete with the brand name product. Like the money Washington pays farmers not to plant crops, this market rigging keeps prices to the consumer sky high.

The deals these drug companies have cut are eye-popping!

KEEPING GENERICS OFF THE MARKET: HOW THE DRUG COMPANIES FLEECE U.S.

- They paid $136 million to keep four generic companies from making low-cost medicines to compete with the sleep-disorder drug Provigil until 2011.

- Bristol-Myers Squibb and Sanofi-Aventis paid a generic company to refrain from competing with Plavix, the blood thinner, until 2011.

- Schering-Plough and Upsher-Smith Laboratories got a generic drug maker to drop its patent challenge to K-Dur 20, a blood pressure medicine, so they could keep the generic off the market.[53]

But the Federal Trade Commission (FTC) intervened in the 1990s and blocked these agreements, calling them illegal deals in restraint of trade. The drug companies sued, and in 2005 two federal appellate courts overruled the FTC and stopped the commission from blocking these deals. The result, according to FTC Commissioner Jon Leibowitz, gave the drug companies "carte blanche to avoid competition and share resulting profits." He added: "Until recently, payments by brand-name companies to generics were the exception, but now they're the rule. They appear to be a new way to do business, and that's very troubling."[54]

The FTC is planning an appeal to the United States Supreme Court to allow it to continue blocking these deals to keep generic drugs off the market.

When the pharmaceutical companies cannot buy off their competition, they sometimes resort to smear campaigns to discredit them. *New York Times* op-ed contributor Daniel Carlat, a professor at Tufts University School of Medicine and the editor in chief of *The Carlat Psychiatry Report,* revealed a campaign by the drug companies Sepracor, Sanofi-Aventis, and Takeda to besmirch the reputation of a competitor sleep aide, trazodone.[55]

The market for sleeping pills is huge—42 million prescriptions were filled last year—and more competitive than ever, thanks to the recent introduction of Sepracor's Lunesta (the one with the butterfly commercials), Sanofi-Aventis's Ambien CR (an extended-release version of Ambien), and Takeda Pharmaceuticals' Rozerem. Ads have made most of these drugs household names. Yet many people have never heard of one of the most widely prescribed hypnotics in the United States: trazodone.

Approved twenty-five years ago by the FDA, trazodone is available in generic form and costs only about ten cents per pill compared with Ambien and Lunesta, whose cost can range up to three dollars and even higher.

Dr. Carlat explains that trazodone "is categorized as an antidepressant. Nonetheless, psychiatrists prescribe it off label to treat insomnia, because it works so well. Trazodone carries no risk of addiction; its half-life is long enough to keep patients asleep all night. . . . And in the only sizable study to compare trazodone with Ambien as a sleep aid, the two drugs performed equally well." [56]

"In the past few years," Carlat reports, "several articles have been published in professional journals that can only be described as trazodone-bashing." But, he underscores that the "authors, psychiatrists with university affiliations, have been paid by Sepracor, Sanofi-Aventis or Takeda, the companies that stand to gain from trazodone's downfall." [57]

The articles criticize trazodone for having side effects "like cardiac arrhythmias or priapism (prolonged, painful erections). But these side effects are extremely rare: priapism has been found to occur in one in five thousand men who take the drug, and the incidence of cardiac arrhythmias is even lower." [58]

Apparently, misleading negative advertising isn't confined to electoral politics!

Recently, big drug companies have begun trying yet another tactic in their war against generics. When the patent on their high priced drugs runs out, many pharmaceutical companies are launching their own generic substitutes, undercutting the price the independent generic drugmaker might charge. While this would seem to be a modern application of the old adage, If you can't beat 'em, join 'em, the drug companies are likely only lowering their prices temporarily, just long enough to break their competitors. After they vanquish the competition, the companies can safely charge sky-high prices again. Senator Chuck Schumer (D-NY) called this a form of "predatory pricing" and accused the drug companies of "trying to put a dagger through the heart of the generic industry." [59]

For example, in June 2006, Merck's patent on its anti-cholesterol drug, Zocor, expired. As the New York Times reported, the med had earned the company $4.4 billion in revenues. [60] But generic competition was a serious threat: the Times reported that Zocor was soon to be the "largest-selling drug yet to be opened to generic competition." The newspaper noted that Teva Pharmaceutical Industries, the Israeli generic drug company, had an exclusive 180-day window to compete with Zocor. Instead of true exclusivity, however—with its attendant profits—Teva found itself facing competition from a laboratory licensed by Merck to manufacture generic Zocor. Dropping the generic price to its rock-bottom level would help Merck make

Teva's generic unprofitable, discouraging others from entering the generic business.[61]

The Justice Department has yet to comment on whether the pharmaceutical company's conduct violates antitrust laws, but it would seem to be a prototypical example of monopolistic pricing to create and preserve a future monopoly. Kathleen Jaeger, the president of the Generic Pharmaceutical Association, calls it—quite rightly, we believe—"the new scam in town."[62]

This combination of lobbying, spending money to defeat ballot propositions, buying off generic competitors, and smearing the reputation of cheaper generic drugs is keeping in place a gigantic profit center for the huge drug companies—all at our expense. Whether we pay these extra costs over the counter or in our insurance premiums or in our tax bills for Medicare and Medicaid, we consumers surely pay for them.

And it's an outrage!

ACTION AGENDA

The most effective way we can limit the drug industry's power to push prescriptions and overmedicate America is by working with our individual doctors to battle their influence. When you go for medical treatment for yourself or for a member of your family, you can take some simple steps to protect yourself:

- Always ask your physician or surgeon if there is a generic drug that could be substituted for the prescription medication he or she is proposing.

- Ask if your doctor has any financial relationship with any drug company.

- Share with your doctor your opinion about these "relationships" to let him or her know that you disapprove of these frequently shady transactions.

Most doctors pride themselves on their ethics and integrity. They endure long hours and frequently treat nonpaying patients because of their innate idealism. But they have been seduced by the pharmaceutical industry, which has gradually persuaded many practitioners that they can be "a little bit pregnant"—that meals or gifts from sales reps don't compromise their independence. You need to press your health care provider in the opposite direction. Make them feel their patients' discontent over cozy relationships with sales reps.

Politically, conservatives and Republicans who espouse lower government spending must come to see these drug companies' efforts to drive up federal, state, and local spending for the egregious economic sinkholes they are. The single biggest thing that we, as taxpayers, can do to cut government spending is to hold down Medicare and Medicaid costs. This is not an issue of intrusive government regulation, but rather one of stopping wasteful public spending. If people want to be overcharged for medication that they pay for themselves, or through insurance, that's their business. But when the drug companies fleece the taxpayers by scamming Medicare and Medicaid and by outfoxing the makers of generic drugs, that's a public issue for fiscal conservatives!

Work with patient advocacy groups in your state to promote ballot initiatives requiring drug companies to offer discounts to Medicaid patients. The defeat in California shouldn't stop others from trying. Once a few states adopt these measures, they'll catch on throughout the country. And we may even force Washington to act!

FANNIE MAE:
The Democratic Enron

If there is such a thing as a "vast left wing conspiracy," its financial epicenter is an organization named after a girl. Established as part of the New Deal as a government agency, the Federal National Mortgage Association spun off in 1968 to become a semi-independent entity known as Fannie Mae.

Its nominal purpose is to enhance mortgage lending and home owner-ship by purchasing mortgages, lent by banks and other lending institutions, and selling mortgage-backed securities on the financial markets to generate additional capital to buy up more mortgages. This secondary mortgage market frees banks to keep lending money for new homes even after they have exhausted their capital. Before interstate banking, liquidity was a seri-ous problem for local banks, savings and loan associations, and credit unions. Fannie Mae succeeded in providing liquidity so the flow of mort-gage money could continue.

But Fannie Mae's mission has spun off to create a massive, quasi-governmental entity. Underpinning Fannie Mae is an implicit, although not legally articulated, guarantee that the federal government will come to its rescue should it run into financial trouble. This understood assurance allows Fannie Mae to sell its bonds at a lower rate than private lenders. In addition, Fannie gets other public benefits, such as exemption from certain disclosure requirements, regulatory review, and important tax exemptions.

The Treasury also is authorized to purchase up to $2.25 billion of Fannie mortgages to preserve the organization's liquidity.

But the unintended consequence of Fannie Mae's quasi-public status has been to build an empire of the Left—a self-perpetuating private company that uses public benefits to maintain a large patronage and funding operation that sustains liberal causes throughout the nation.

The green Elysian fields of Fannie Mae offer top bonuses to key Democrats. This was all well and good, until federal regulators discovered a nasty secret: Fannie Mae had been cooking the books—overstating its profits by $9 billion, which jacked up the bonuses its favored Democrats were entitled to receive.

Fannie Mae is to the Democrats what Enron was to the Republicans—the ultimate example of corporate greed and scandal.

THE NEW DADDY WARBUCKS:

Franklin Delano Raines

- As CEO of Fannie Mae, "earned" $90 million in compensation over six years
- As director of Bill Clinton's Office of Management and Budget (OMB), stressed frugality!

The top beneficiary was Franklin Delano Raines, a former director of the Office of Management and Budget (OMB), who raked in astonishing personal income during his six-year tenure. In 2002 alone, Fannie Mae's board of directors raised Raines's compensation by 50 percent, to $17.1 million—including $200,000 in private jet travel and $37,000 in tax counseling and accounting services.[1]

But others were in on the feeding frenzy as well.

ON THE GRAVY TRAIN:

The Other Deserving Democrats[2]

- Former Mondale campaign manager Jim Johnson, who piloted his man to the fiasco of 1984, went on to his reward as CEO of Fannie Mae before Raines took the job. Johnson's most recent task: handling the vice presidential search for John Kerry in the 2004 campaign. His compensation: $2.9 million!

- Tom Donilon, who coached Mike Dukakis in 1988 and Bill Clinton in 1992 for their presidential debates, became executive vice president of Fannie Mae.

- Jamie Gorelick, the former Clinton administration assistant attorney general (and Janet Reno's designated handler), made $25 million in salaries and bonuses after becoming vice chairman at Fannie Mae! Gorelick also served as Clinton's unofficial representative on the 9/11 Commission: It was the famous "Gorelick Wall" between intelligence agents and investigators that had a lot to do with stopping the FBI from finding out about 9/11 before it happened.[3]

- Bill Daley, brother of the mayor of Chicago, secretary of commerce in the Clinton administration, and Gore spokesman during the postelection controversy in 2000, is another favored Democrat who got a seat on Fannie's board.

- Jack Quinn, Al Gore's vice presidential chief of staff and Clinton's compliant White House counsel, was also rewarded with a board seat. (Quinn was also the lawyer who helped fugitive Marc Rich secure a pardon.)

Republicans have also lined up at the trough, although they're farther back in line:

- Robert Zoellick, who prepped George W. Bush for his presidential debate, is a board member.

- Chuck Greener, former communications director at the Republican National Committee, got another seat.

- Ken Duberstein, Reagan's chief of staff, serves on the board.

- Ann McLaughlin Korologos, Reagan's labor secretary, is on as well.[4]

Fortune magazine reports that Fannie Mae paid its twenty most senior executives a total of $245 million in bonuses over five years; in 2002, each of its twenty-one highest-ranking execs collected more than $1 million.[5]

But the greed of these former politicians for bonuses and exorbitant salaries is only the beginning of the story. The fact is that their juggling of the books at Fannie Mae could still end up costing the taxpayers hundreds of billions of dollars. As a public-private company, Fannie Mae chooses most of its board (though the president appoints five of the thirteen board members), so the company could do as it pleased without federal control. But it still keeps its government benefits.

FANNIE MAE'S GOOD DEAL

- Low interest rates because of implicit federal guarantee
- Exempt from SEC regulation
- No state or local income taxes
- U.S. Treasury buys their debt
- Less stringent capital requirements

Fannie Mae pays one of the lowest interest rates in the nation, only a touch lower than the Treasury Department itself. Why? It issues bonds that everyone assumes are federally guaranteed, although legally they haven't been since Fannie Mae went private. But investors paid no attention. They remain sure that Congress can never allow Fannie Mae to fail—and they're right.

Fannie Mae is exempt from Securities and Exchange Commission (SEC) rules. They don't have to "describe their bonds and securities to prospective investors or pay fees to register them—costing the government millions . . . a year. Nor have they had to file annual reports with the SEC, notify the public of major changes in their financial condition, or disclose when executives buy or sell company stock." Fannie Mae says it sometimes makes these disclosures voluntarily, but Congressman Richard Baker

(R-LA), one of the good guys and chairman of the House committee that oversaw Fannie Mae in the Republican Congress, says "if those disclosures are voluntary, bank robbery is a charitable activity."[6]

Fannie Mae also gets away without paying any income taxes to state or local government—even to the District of Columbia, where it is headquartered. Said former D.C. council member Bill Lightfoot: "We have so many people in the District going without, while Fannie Mae sits rich, fat, and happy with its huge salaries and profits. We could fix a lot of dilapidated schools and buy a lot of fire trucks if they paid their fair share."[7]

How much do these government benefits cost the taxpayer? To answer that question, the Congressional Budget Office asked how much a regular private company would have to pay that Fannie Mae does not because of its special treatment. The answer? "The subsidy to Fannie Mae [and its smaller competing stepbrother] Freddie Mac is worth about $11 billion a year."[8] Both Alan Greenspan, former chairman of the Federal Reserve Board, and the U.S. Department of the Treasury have endorsed the Congressional Budget Office's findings.

How much of that benefit does Fannie Mae pass along to the consumer? The budget office found that "for every three dollars of the subsidy that the company passes on to homeowners, it keeps almost another $2 for its stockholders and executives."[9]

That means we taxpayers are subsidizing more than $4 billion in payments to stockholders and executives of Fannie Mae each year—including the deserving Democrats it employs.

Ever since FDR set up Fannie Mae to stimulate home construction during the Great Depression, the agency has trumpeted its role in promoting mortgage liquidity. It was one of the pioneers in marketing mortgage-backed securities, which offered investors a way to realize the higher returns that mortgages offered without incurring any real extra risk, and also injected added funds into mortgage lending.

Since it was spun off from government control in 1968, Fannie Mae has grown huge. It says that it services 40 percent of the home mortgages in the United States. Fannie Mae bought or guarantees $600 billion in mortgages every year and makes an annual profit of $6 billion.[10]

Its erstwhile chairman, Frank Raines, was blunt in talking about Fannie Mae's public generosity. "Our Mission is to tear down barriers, lower costs, and increase the opportunities for homeownership and affordable rental housing for all Americans. Because having a safe place to call home strengthens families, communities, and our nation as a whole." [11]

So why should Fannie Mae continue as a government-sponsored company? Why not privatize it completely and curtail the subsidies it gets? "It's time to say, 'Thanks guys, you've done well,'" says NYU professor Lawrence White. " 'But now you can go home and swim like everyone else.'" White wants to tailor a program more narrowly to meet the needs of poor people, particularly potential minority borrowers. [12]

With the average Fannie Mae loan coming in at $118,000, many people question whether it successfully serves the needs of the poor. William Apgar, a top official in Bill Clinton's Department of Housing and Urban Development (HUD) and now a professor at the Kennedy School of Government, makes the case that Fannie Mae actually *decreases* the ability of African Americans and Hispanics to buy homes. [13]

Ross Guberman, reporting for *Washingtonian* online, writes that, whatever Fannie Mae's and Freddie Mac's good intentions, "for years HUD has found that Fannie and Freddie lag behind private lenders in serving African American borrowers and other underserved communities." [14]

But, of course, the biggest government subsidy to Fannie Mae and Freddie Mac is the one the government does not yet have to pay . . . but may yet have to. Should either run into adversity, the inevitable taxpayer bailout would dwarf even the massive relief granted depositors in savings and loan associations in the 1980s.

Fannie Mae has needed federal help before. It lost hundreds of millions of dollars in the early 1980s, when interest rates skyrocketed, and had to rely on tax relief and other bailouts from Congress. The Farm Credit System, a parallel effort to Fannie Mae, was similarly bailed out a few years later.

As Guberman reports, "Together, Fannie Mae and Freddie Mac have a trillion and a half dollars in assets—more than the gross domestic product of every country but Japan, Germany, and the United States. Their combined debt . . . is poised to outpace that of the U.S. government." Eight trillion dollars flows through its bank account every year. [15] It's a precarious situation, as House Oversight Subcommittee Chairman Richard Baker (R-LA) says, chillingly: "The taxpayers are living under an enormous rock suspended by a single rope. Once it breaks, there's no recovery." [16]

A lot of people are concerned that the risk of a Fannie Mae collapse is more than just academic speculation. The *Wall Street Journal* has editorialized about the parallels between Fannie Mae and Enron, calling attention to their debt levels and "terrible" record of financial disclosures. The *Economist* wrote that Fannie Mae and Freddie Mac are "arguably the most worrying concentrations of risk in the global financial system." [17]

The fact is, if interest rates rise to sky-high levels, as they did in the early 1980s, or property values collapse under a glut of supply, Fannie Mae could be in trouble—and the taxpayers would be left to carry its weight.

And the risk is more than theoretical, because behind Fannie Mae's closed doors, the company has been engaging in exactly the conduct that brought down Enron: overstatement of earnings and profits. The higher earnings rose, the more salary, bonuses, and other compensation the Fannie Mae bigwigs got!

Here is where Frank Raines, Clinton's Director of the Office of Management and Budget, really excelled. His scheming started when he persuaded the Fannie Mae board to tie his compensation to earnings. The mission of Fannie Mae was always supposed to be to help underserved people buy homes, not to make money. But Raines changed the name of the game.

According to Bethany McLean of *Fortune* (who is credited with breaking the Enron story), "the fact that Frank Raines's compensation was based mostly on meeting earnings per share goals and not on doing what the company was supposed to do for affordable housing, I think means that you have to look somewhat askance at it." [18]

So how did Raines manage to get the Fannie Mae board to vote him a compensation arrangement that ran to more than $15 million per year? He made sure they also got access to the cookie jar.

But he did more than that. He also arranged to funnel Fannie Mae funds to their favorite charities. The Fannie Mae Foundation, which was giving out almost $50 million annually in grants, began to give priority to the favored charities of those whose votes Raines needed—his board of directors. As the Associated Press reports, a lawsuit filed by Fannie Mae shareholders on February 27, 2006, accused Raines of "us[ing] charitable contributions to compromise the judgment of six board members." [19] The suit alleges that he used Fannie Mae's charitable foundation, which he chaired, to pour "millions of dollars in grants into organizations in which other members of

Fannie Mae's Board held senior positions." The shareholders said that the donations made the directors "beholden" to Raines, and was the reason that "they never raised any opposition to the CEO during dozens of board, audit committee, and compensation committee meetings 'enabling and fostering the perpetuation of his schemes.'" [20]

The donations, which the suit says totaled $1.75 million, included contributions to Johns Hopkins University, Howard University, and the National Alliance to End Homelessness, Inc., "organizations whose board members included Fannie Mae directors." [21]

Raines also used dreams of empire building to hold his board in line. He promised Fannie Mae's shareholders a 15 percent growth in earnings each year. But the *Washingtonian* estimates that mortgages, the basis for Fannie Mae's operations, are only expected to grow at half that projected rate. [22]

What to do? Raines decided to expand beyond the secondary mortgage business into direct lending. As Guberman reported in the *Washingtonian,* "Fannie has tried to sell foreclosed homes directly and tried to provide life insurance to homebuyers as a 'public service'—until someone leaked a memo about how the program would increase company profits. . . . Fannie [also] partnered with Home Depot's upscale Expo stores to provide installment loans—until critics scoffed that taxpayers shouldn't be subsidizing hot tubs and sunrooms." [23]

Private lenders, who are not endowed with Fannie Mae's government largesse, worry that the agency is trying to move toward making direct mortgage loans. As long as Fannie Mae and Freddie Mac only buy mortgages that have already been lent by a bank or other credit institution, private lenders can live with it. But they fear a government-subsidized monolith squeezing them out of the mortgage-lending market. As Guberman notes, Morgan Stanley analyst Ken Posner has privately forecast that Fannie Mae's unique advantages and management greed would soon start driving lenders out of business. [24]

When Raines saw that he couldn't keep his promises to his shareholders by diversifying into other businesses, he needed a new way to increase corporate profits to a point where he could cash in and get a huge bonus. So he hit on a solution: cook the books.

The stakes for Raines couldn't have been higher. Of the $90 million he made in compensation as CEO of Fannie Mae, $52 million had been tied to the company's earnings. To rake in that kind of money, he had to show huge and skyrocketing profits, which were simply impossible. So he embarked on what federal regulators have called a massive effort in which "senior management manipulated accounting, reaped maximum, undeserved bonuses and prevented the rest of the world from knowing." [25] Raines, who still may face criminal charges, denies any knowledge of shady accounting practices. But a report by the Office of Federal Housing Enterprise Oversight (OFHEO), the body charged with regulating Fannie Mae and Freddie Mac, "blasts [Raines] for his role in creating a corporate culture that let senior management earn huge bonuses based on earnings and manipulated those earnings." [26]

The OFHEO attacked what it said was an "arrogant and unethical corporate culture," and that management had "stonewalled" its inquiry. It urged that Fannie Mae investigate "employees associated with the accounting scandal" further. [27]

For his part, Raines was forced to resign on December 21, 2004.

How Did Fannie Mae Get Away With It?

The question that lingers is how Fannie Mae managed to get away with its outrageous conduct for so long. How does the CEO of a government-linked institution get away with making $90 million? And how has Fannie Mae been able to keep its government benefits while embarking on so obviously capitalistic a course?

The key is lobbying. Raines's Fannie Mae was a power to be reckoned with on Capitol Hill. It offered luscious carrots and wielded an enormous stick.

Fannie Mae hyped its public image by spending $44 million annually on television advertising. Why would a company that makes no direct loans to consumers—and deals only in the secondary mortgage market, where it buys loans made by banks and other lenders—need to advertise? Fannie executives claimed it was part of a program of public outreach. More likely, it was designed to develop and magnify Fannie Mae's political clout to protect its government privileges.

Raines assured his political strength by employing some of Washington's top lobbyists and political powerhouses. In its heyday, Fannie Mae spent $6 million annually on political public relations and consultation from Washington heavyweights.

Consumer advocate Ralph Nader has called this "Rolodex hiring," charging that "it's all a matter of know-who, not know-how. They've perfected all the techniques of lobbying and pay massive salaries . . . to ensure against any change" in their government benefits.[28]

When hit with negative media coverage, Raines's crew responded like the Clinton War Room with defense, rebuttal, counterattack, and—where they would work—charm and payoffs. According to the *Washingtonian*, when the *Wall Street Journal* ran a story critical of Fannie Mae's practices, the organization "called the editorials a 'smear job' and orchestrated protest calls. Top [Fannie] executives Arne Christenson and Chuck Greener—who knew editorial page editor Paul Gigot from conservative circles—traveled to Manhattan to mend fences." [29]

Susan Lee, the author of the critical editorials, said "after sending around all these nasty letters and launching these incredible missiles, they tried to jolly us, Republican to Republican." Christenson passed around old photos of Gigot and himself to underscore his connection to Lee's boss.[30]

Part of Fannie's "Rolodex hiring" strategy was designed to neutralize opponents. When House Republican Richard Baker (R-LA), then chairman of the House Financial Services Subcommittee on Capital Markets, Insurance, and Government Sponsored Enterprises, breathed too heavily down Fannie Mae's neck, the company hired his chief of staff, Duane Duncan, to work for them.

When charm failed, though, Raines wasn't afraid to throw his weight around. After Congressman Baker proposed a stronger regulator for Fannie Mae than the seemingly obliging OFHEO, the company "hired a phone bank to call [his] constituents on behalf of the Coalition of Preserve Homeownership, a front for the interest of Fannie [Mae], Freddie [Mac], real-estate agents, and homebuilders. Some members of Baker's subcommittee were enraged after they received anonymous boxes filled with thousands of letters from constituents protesting a so-called congressional proposal 'to raise mortgage costs.' " [31]

In a particularly outrageous lobbying act by this government-linked company, Fannie Mae ran a television advertisement on March 31, 2004, the day before the Senate Banking Committee began to work on regulatory legislation. The ad text, featuring a Hispanic couple, was an outrageous scare tactic.

FANNIE MAE'S SHAMELESS AD

Man: "Uh-oh."

Woman: "What?"

Man: "It looks like Congress is talking about new regulations for Fannie Mae."

Woman: "Will that keep us from getting that lower mortgage rate?"

Man: "Some economists say rates may go up."

Woman: "But that could mean we won't be able to afford the new house."

Man: "I know." [32]

Senator Chuck Hagel (R-NE) chastised Fannie Mae for the ad: "Here is an organization that was created by the Congress . . . spending money questioning the Congress's right to take a serious look at oversight," he sputtered during the hearing that day. "I find it astounding. Astounding!" [33]

Raines also used the Fannie Mae Foundation to great effect in enlisting charitable enterprises, recipients of his grants, to argue his case before the public. Nader, who calls Fannie Mae's charitable giving "grant payola," points out that "Fannie sprinkles millions around the District [of Columbia] and then calls on those groups when the company needs to neutralize dissent." [34]

Fortune reports that the Fannie Mae "foundation had existed in a small form since 1979, but in 1996, Fannie Mae seeded it with $350 million of its own stock and gave it responsibility for Fannie's advertising. Over the past five years the foundation has given away some $500 million to thousands of organizations ranging from the Congressional Black Caucus to the Cold Climate Housing Research Center in Fairbanks [Alaska]. [35]

In 2004, for example, the Fannie Mae Foundation reported assets of $179 million, and awarded grants that year totaling $48 million. The foundation said that its focus is expanding home ownership, particularly for minority and low income families, and notes that by January 2005 it had provided 17 million people with "free, step-by-step home-buying information to help them achieve the American dream of home ownership." [36]

But Fannie Mae's real role seems to be as a retirement home for deserving Democratic politicians, and a funding source for virtually every liberal group in Washington. Among the organizations this quasi-governmental agency's foundation has given to are a who's-who of the Left.

WHO GOT THE MONEY?

ORGANIZATIONS THAT GOT GRANTS FROM FANNIE MAE'S FOUNDATION [37]

- The National Council of La Raza
- The National Association for the Advancement of Colored People Legal Defense & Education Fund
- The Mexican American Legal Defense & Education Fund
- The Association of Community Organizations for Reform Now
- The Children's Defense Fund
- The National Urban League
- The Center for Community Change
- The American Civil Liberties Union Fund of the National Capital Area
- The Brookings Institution
- The Urban Institute
- Global Rights
- Women Empowered Against Violence
- The National Organization for Women Legal Defense and Education Fund
- Alliance for Justice
- The National Committee for Responsive Philanthropy
- Planned Parenthood
- The Institute for Policy Studies
- The National Council of Negro Women
- The National Black Caucus of State Legislators

- The National Women's Law Center
- The Lawyers Committee for Civil Rights Under Law
- The National Alliance to End Homelessness
- The Organization for a New Equality
- The Congressional Black Caucus Foundation
- The Southern Christian Leadership Conference
- The National Association of Latino Elected Officials
- The National Conference for Community and Justice
- Mi Casa My House
- The Center for Policy Alternatives
- The Multicultural Career Intern Program
- The Washington Legal Clinic for the Homeless
- The National Immigration Forum
- The Congressional Hispanic Caucus Institute
- The African American Institute
- The Latin American Association
- WAGES International—Women's Alliance
- The National Organization on Disability
- The Women's Legal Defense Fund
- The Gay & Lesbian Community Action Council
- The International Human Rights Law Group
- The Legal Aid Society
- The National Association for Public Interest Law
- The National Legal Aid & Defender Association
- The National Political Congress of Black Women
- The Conservation Law Foundation
- The New York Immigration Committee
- The Progress and Freedom Foundation
- The Center to Prevent Handgun Violence
- The Alliance for Fairness in Reforms to Medicaid

(continued)

- The Green Institute
- Parents, Families, & Friends of Lesbians & Gays
- The Council of Latino Agencies
- The Citizenship Education Fund
- The District of Columbia Appleseed Center for Law and Justice
- The Empowerment Network Foundation
- The Conservation Fund
- Catalyst for Women
- Demos: A Network for Ideas and Action
- The African American Women's Resource Center
- The AIDS Action Foundation

"Fannie Mae is a bit like the tobacco industry," says Charles Lewis, executive director of the Public Interest Research Group (PIRG) for Washington, D.C. "They use the foundation to try to put a happy face on what they do. The whole thing is part of a perpetual boondoggle so they maximize public risk and private gain."[38]

For example, after Connecticut congressman Christopher Shays, a progressive Republican, criticized Fannie Mae's failure to disclose the same financial information as the SEC required of other private corporations, he suddenly started receiving letters opposing his proposal from the National Urban League and the National Council of La Raza, a Hispanic advocacy organization—letters Fannie Mae probably stimulated.[39]

And when John Taylor, president of the National Community Reinvestment Coalition, a group representing hundreds of local community groups, complained that Fannie Mae was not doing enough for affordable housing, he lost his $65,000 grant (though he later got it back). Reverend Graylan S. Hagler, pastor of the Plymouth Congregational Church of Christ in Washington, who criticized Fannie Mae's lending practices, was punished when Fannie's foundation cancelled an $80,000 grant to a nonprofit housing group he supported. The *Washingtonian* reports that "he picketed Fannie's headquarters and its annual meeting in Texas in response."[40]

And, of course, Fannie Mae's PAC and executives give massive campaign contributions. Congressman Ken Bentsen (D-TX), son of the former Texas senator Lloyd Bentsen and a member of Congressman Richard Baker's oversight subcommittee, got eleven checks from Fannie Mae employees in October 2000 as he ran for office. Senator Chuck Robb (D-VA), on the Senate Banking Committee, got fourteen checks from Fannie Mae employees.[41]

Fannie Mae also exerts influence through its "partnership offices," local Fannie Mae outlets that are supposed to target innovative mortgage products in particular neighborhoods.[42] Sometimes, however, these seem to be no more than political clubs to promote Fannie Mae. These "partnership offices" are often staffed by former politicians like Bob Simpson, former aide to Democratic Senate leader Tom Daschle, who heads the South Dakota office. One of the first of these to open, oddly, was in the district of the then chairman of the House Banking Committee, Congressman Henry Gonzalez of San Antonio, Texas.

The net effect of all this lobbying has been to lend Fannie Mae a sense of tremendous power. Former Housing and Urban Development Secretary Andrew Cuomo has observed (bravely), "You did not question Fannie Mae. Fannie did as Fannie wanted."[43]

Indeed, Fannie Mae is quick to jump all over anyone who questions it. As CNN has noted, "in 1996, when the Congressional Budget Office issued a critical study, a Fannie spokesperson sneered that the report was 'the work of economic pencil brains who wouldn't recognize something that works for ordinary homebuyers if it hit them in their erasers.' And it was quick to remind politicians where their interests lay. Fannie put together a book, personalized for each member of the House Banking Committee, detailing all the good things it did in each district to bolster home-ownership."[44]

But Fannie's biggest fight—before the accounting scandal laid it low—was against the Bush administration. In September 2003, Treasury secretary John Snow and HUD chief Mel Martinez testified before the House Financial Services Committee. Snow "called for a new regulatory regime, one key element of which would be a receivership program," CNN reported. "By setting up a receivership mechanism, the government would be sending an unmistakable message: it would not stand behind Fannie and Freddie's debt."[45]

That meant war.

Legislation to enact the administration proposals was going through the Republican-controlled Senate until Senator Bob Bennett of Utah—whose son happened to be deputy director of Fannie's Partnership Office in Utah—effectively killed the bill with the help of a number of other Republicans.

After the rebuke, Raines sent Snow an angry letter charging him with "disrupting the capital markets by indicating a wish to change Fannie Mae's charter, status, or mission." Because Snow had dared to propose such a change, he scolded, "trading in our debt came to a halt for an extended period of time. I am disappointed and hope we can change course."[46] The message: Just because we at Fannie Mae expect and get an implicit guarantee of federal subsidy should we need it, and just because we use that status irresponsibly by inflating our earnings and drawing huge salaries and bonuses and giving away tens of millions to every liberal group we can find, how dare you criticize us? How dare you rain on our parade?

Raines Gets What's Coming to Him . . . Up to A Point

The fall, when it came, was sudden and sharp. And it left Raines and his liberal clique in ruins.

It began on September 22, 2004, when Fannie's overseer, OFHEO, released a two-hundred-page "special examination of Fannie Mae," saying that the foundation was "both willfully breaking accounting rules and fostering an environment of 'weak or nonexistent' internal controls." In particular, OFHEO "focused on . . . Fannie's use of accounting rules to defer derivative losses onto its balance sheet . . . OFHEO said that Fannie hadn't just bent the rules, it had broken them."[47]

OFHEO charged that a $200 million understatement of expenses "allowed the company to report earnings of $3.23 per share, which meant that Fannie paid out a total of about $27 million in bonuses."[48]

Raines quickly attacked the report. "These accounting standards are highly complex and require determinations on which experts often disagree," he said, adding that "there were no facts" to support the charge that Fannie executives had deferred an expense to earn more bonus money.[49]

And the Democrats rallied round. "This hearing is about the political lynching of Franklin Raines," said Congressman William Lacy Clay

(D-MO). Congressman Barney Frank (D-MA) agreed: "I see nothing in here that suggests that safety and soundness are an issue."[50]

"Is it possible that by casting all of these aspersions . . . you potentially are weakening this institution in the market, that you are potentially weakening the housing market in this country?" Alabama congressman Arthur Davis asked. "As the hearing continued," according to Bethany McLean, "Davis acted like a prosecutor grilling a hostile witness. He wanted a one-word answer: yes or no."[51]

As McLean notes, Frank Raines then "overplayed his hand one last time." He demanded that the Securities and Exchange Commission (SEC) review OFHEO's findings. Raines must have been confident of success when he had Fannie once again reach for the extra edge of influence: he hired Bill McLucas, of the law firm Wilmer Cutler, to make the case before the SEC. McLucas, it turned out, had formerly been the SEC's chief enforcement officer.[52]

McLean describes Raines's final hour: "At a little after 6:00 PM on December 15, 2004, about thirty people—including Raines . . . piled into a conference room at the SEC's headquarters in Washington, D.C., and seated themselves around a large rectangular table. SEC officials made it clear that they had not addressed any issues of individual culpability—those investigations were ongoing. The chief accountant, Donald Nicolaisen, announced that the SEC had decided that Fannie did not comply 'in material respects' with accounting rules and that, as a result, Fannie would have to restate its results."[53]

"Raines looked stricken . . . It was over for Frank Raines."[54]

The Fannie Mae CEO quit on December 21, 2004. But he is still not out of the woods. Criminal investigations by the Justice Department are continuing.

The OFHEO report, backed up by the SEC, says that "Fannie employees manipulated earnings to trigger bonuses for management from 1998 to 2004." Acting OFHEO director James Lockhart said "the image of Fannie Mae as one of the lowest-risk and 'best in class' institutions was a facade. Senior management manipulated accounting, reaped maximum, undeserved bonuses, and prevented the rest of the world from knowing."[55]

Fannie Mae agreed to pay a $400 million fine, most of which will go back to investors, according to the SEC. It will have to implement thirteen recommendations made by the OFHEO, including using independent consultants to verify compliance with the regulator's rules, developing new guidelines for its board, and a "review" of people who are mentioned in the [OFHEO] report, including the new CEO, Daniel Mudd.[56]

The biggest change is that Fannie Mae must cap the growth of its portfolio and roll it back to the $727 billion it held as of the end of 2005. Federal regulators worry that, with Fannie Mae and Freddie Mac holding a combined total of $1.4 trillion in mortgage-backed securities, a failure could put the entire national economy in danger.

Has Fannie Mae learned its lesson? The company's new CEO, Dan Mudd, is talking a good game: "To the extent that this company strays from its mission of financing affordable housing, we start to lose our compass. . . . There are areas in which we have gotten arrogant."[57]

"We have all learned some powerful lessons here about getting things right and about hubris and humility," Mudd said in a statement.[58]

But Lockhart and SEC Chairman Christopher Cox stress that the investigation of individuals at Fannie Mae is continuing—including further inquiries concerning both Raines and the foundation's current CEO. "Mudd's not out of the woods on this thing," said Paul Miller of Friedman, Billings, Ramsey.[59]

ACTION AGENDA

In the entire investigation of Fannie Mae, investigators largely overlooked its political role. The inquiry, and the resulting reforms, have focused almost exclusively on the financial issues of manipulated earnings reports and inflated bonuses to executives. The solutions have largely involved cleaning up the firm's accounting practices and limiting potential government exposure in the event the company should run into trouble.

But is it really desirable to have a political club controlling such massive assets and profiting from such public largesse? Fannie Mae's foundation is an essential element of the support structure of liberal GSEs (government-sponsored enterprises). Distributing nearly $50 million each year—all to left-wing groups—is a vital part of their financial solvency. Fannie Mae's employment of top politicians, mostly Democrats, has created a kind of retirement home for the Left.

If these were the activities of a group like MoveOn.org or some other liberal organization, it would be understandable and beyond reproach. But here a quasi-public agency, which survives because of its government benefits and the implicit public guarantee of its bonds, is overtly operating as a political organization, funding its allies, punishing its tormentors, and hiring party bigwigs.

Fannie Mae must be depoliticized. Its foundation should be required to restrict its grants to non-political, housing-oriented entities. The agency should be restricted from hiring every retired Democratic hack it can find. And the company's publicity and lobbying apparatus must be dismantled.

In an ideal world, we might insist that Fannie Mae just sever its ties with the government and operate as a wholly private entity. But its portfolio of secondary mortgages is too large to permit it to operate without some sort of implicit government guarantee. Since Fannie Mae is too big to be floating around in the private sector, it must be reformed to purge it of its political bias and make it a nonpolitical finance agency—as Congress originally intended it to be.

THE BANKRUPTCY BILL:
"To Hell With Forgiveness," Saith the Congress

Wonder why people hated the Republican congressional majority enough to vote them out of office in the 2006 elections? Here's a good example:

In 2005, impelled by massive campaign contributions from banks and credit card companies, Congress passed the first change in the bankruptcy law in twenty-seven years. A draconian bill, it ends the concept of a "clean start," which has long animated the humanitarian impulse behind the bankruptcy process.

A more cold-blooded piece of legislation would be hard to envision. Credit card companies, the beneficiaries of the bill, posted a profit of $30 billion in 2004,[1] charging exorbitantly high interest rates, peddling credit cards to teens like drug pushers, and charging loan shark-like excessive penalties for even slightly late payments. To augment their profits, the revised bankruptcy law holds debtors in almost permanent bondage to their debts, making a mockery of the very purpose of the bankruptcy statute in the first place. In its exclusively pro-business attitude, and with complete lack of compassion for debtors, the Republican Congress conformed to the stereotype of heartlessness that Democrats delight in painting. President

George W. Bush's enlightened "compassionate conservatism" is nowhere to be seen in the legislation.

The revival of debtor's prison cannot be far behind!

WHY DO PEOPLE FILE FOR BANKRUPTCY?

BECAUSE THEY HAVE TO!

- Half of all bankruptcies are filed because of high medical bills.
- One in ten people who file for bankruptcy has cancer when they file.
- Ninety-five percent of bankruptcies are due to job loss, family breakup, or medical bills.
- Only 2 percent of filers are in alcohol or drug rehab when they file.

Source: Newsweek [2]

In their lust to appease their campaign contributors, did our congressmen and senators stop to contemplate—as both *Newsweek* and Harvard researchers have—that more than half of all bankruptcy filers cite medical costs as the key factor that pushed them over the financial edge?

The conservative sponsors of the bankruptcy bill were right that filings have skyrocketed in recent years—from 200,000 in 1978 to 1.6 million in 2004.[3] In fact, according to the American Bankruptcy Institute, one in sixty households filed for bankruptcy in 2005.[4] But the real question is: why the increase in bankruptcy? The key factor is the extraordinary increase in credit card debt, which has quadrupled since 1990 and now comes to just under $700 billion. Interest rates on credit card debt are often not capped, and penalties—often triggered by a single late payment—frequently exceed 30 percent.[5]

The rapid fall in health insurance coverage exacerbates the problem. Forty-one percent of moderate and middle-income adults lacked health insurance for part of 2005, up from 28 percent in 2001. According to the Census Bureau, there are 45 million uninsured Americans.[6]

Congress has failed to provide national health insurance, and the states have failed to regulate credit card interest rates and lending practices. The inevitable result is a major increase in bankruptcies. Yet when amendments were offered on the floor of Congress to exempt bankruptcies caused by medical bills from the legislation's requirements, or to cap credit card interest rates at 30 percent, the Republican majorities—intent on passing this purely punitive measure—rejected them out of hand.

The goal of the bankruptcy reform Congress passed was to curtail the use of the "clean slate" bankruptcy offered under Chapter 7 and to force debtors into the more draconian Chapter 13. Under Chapter 13, those in bankruptcy must do their utmost to pay their debts over a five-year period. As an incentive to encourage lenders to accept partial payment, the bankruptcy judge can reduce the debtor's liability by up to 20 percent, but can do no more.

Rather than cure insolvency, Chapter 13 simply perpetuates it. Most people find it impossible to make the required payments over a five-year period, so about 75 percent of Chapter 13 bankruptcies fail. Then, with the debt unforgiven, the hounding by creditors resumes—but this time the debtor has no surcease or shelter until two more years have passed, after which the debtor can refile for bankruptcy and begin the same dismal process all over again.[7]

For most people, Chapter 13 means a life of indebtedness with no relief. One mistake, one serious illness, one lapse into carelessness or irresponsibility, one job loss, and you're serving a life sentence of debt.

On the surface, the criteria Congress has established for entry into the more generous Chapter 7 seem reasonable. Those who earn less than the median income for their state continue to be eligible for Chapter 7 relief under the new law. The median income for a family of four in the United States in 2003 was $65,000 (or about $1,250 per week before taxes), and ranged from $46,000 in New Mexico to $83,000 in Massachusetts.[8]

Those who earn more than the median income cannot enter Chapter 7 unless they can prove that they cannot pay $6,000 of their debt over the next five years. But in calculating their ability to repay their debt, the law uses the IRS standards for allowable expenses: $200 per month for food and $800 for monthly housing costs for a family of four! Given the cost of a Big Mac, it's hard to imagine how anyone could feed four mouths for $6.50 per day.

Essentially, just about anybody who pays federal income taxes is above the median income cutoff in the legislation. They should be able to handle ordinary credit card debt with that kind of income. But they can't handle it if they lose a job or face huge medical bills . . . and Congress has made sure they'll have to live with it for the rest of their lives.

Under the new bill, student loans cannot be forgiven or reduced. The provisions of the old bankruptcy law prevented government-direct or federally guaranteed student loans from being wiped out, but allowed private student loans to be. The new legislation eliminates any such forgiveness.

In computing the value of "collateral" (which includes furniture, clothing, cars, and electronic equipment), the new law specifies that they be assessed at their replacement value—far above the previous law, which directed that they be valued at their current worth. The assumption seems to be that families in bankruptcy will go out on a spending spree and buy new goods rather than make do with what they have. In some states—such as Florida and Texas—that protect homes from seizure by creditors, the new law provides that dwellings worth more than $125,000 cannot be protected unless they were purchased more than three years before filing for bankruptcy.

This is a bill only Uncle Scrooge could love.

Defenders of this iniquitous piece of legislation cite the benefit to other consumers, arguing that those who borrow from credit card companies and then cannot pay their debts drive up the costs for more diligent borrowers, who conscientiously repay their obligations. But with credit card company profits so high, interest so exorbitant, and penalties so frequent and costly, shouldn't we expect the industry to mitigate its greed and help those driven to bankruptcy by such extravagances as chemotherapy for their cancer?

Of course, the revisions have been a boon for some industries. Credit counselors certainly do well by its provisions, since the law requires anyone seeking bankruptcy protection to enter counseling. Legislators had hoped that the counselors could persuade potential bankrupts to sign up for a debt management plan offered by credit counseling services. But a recent study showed that, since the law was passed, out of fifteen thousand people counseled at a typical agency, only 42 (0.3 percent) signed up for such a program.[9]

The only thing the counseling requirement does is assure that agencies offering financial advice get an average of $51 for each session. With upwards of two thousand people filing for bankruptcy daily, this creates a flow of $100,000 per day for credit counseling agencies.[10]

Lawyers also appreciate the reform legislation. The burden of proof on those filing for bankruptcy has at least doubled under the new statute. The American Bankruptcy Association says that under the new law debtors must provide "a list of secured and unsecured creditors; documentation to credit counseling; monthly income and expenses; assets and liabilities; most recent tax return and any earlier returns that were not filed; pay stubs and photo ID." Lawyers—who must now certify that the filings of those seeking bankruptcy are accurate—are liable to sanctions if they are not. All this adds up to much higher legal fees for those who can, obviously, least afford them.[11]

We all know of people who have used the bankruptcy law irresponsibly to evade paying money they legitimately owed. But the changes Congress enacted at the prompting of the industry are truly excessive.

THE CONGRESSIONAL RECORD ON BANKRUPTCY

Global Crossing . . .
Enron . . .
Worldom . . .
         . . . unscathed.
Student loan debtors . . .
Patients in bankruptcy due to medical costs . . .
Broken families . . .
         . . . punished forever!

One does not have to engage in class warfare to ask why Congress was so anxious to crack down on the small individual debtor while leaving unscathed the massive corporate bankruptcies (like Enron, Global Crossing, and WorldCom) that left millions of individuals with sharply depleted investments or without their hard-earned pensions. As the *San Francisco Chronicle* pointed out editorially, the Congress "could have addressed myriad schemes used by the wealthy to shield their assets from creditors. It could have taken on some of the highly profitable but ethically suspect lending practices—predatory mortgage loans to the elderly, high-interest credit cards to college students, payday loans to the working poor—that too often send the recipient into a downward spiral of debt. Instead, the bill focused on an easy target—low-income folks who find themselves in dire straits." [12]

A key opponent of the bill, New York Democratic senator Chuck Schumer (whose irrelevant amendment to stop the law from discharging fines levied by courts against pro-life demonstrators outside abortion clinics held up the bill's passage for more than a year) was particularly critical of the "bias" in the bill. Noting that the law does nothing to close a loophole that allows the rich to shelter assets through trusts, Schumer said the bill "deals with abuses in bankruptcy by one group but not with another group." [13]

At the bidding of the credit card industry, the Republican Congress had tried before to revise the bankruptcy law. It had passed similar legislation in the waning days of the Clinton administration, only to have it die through a presidential pocket veto (in which the president fails to sign a bill within the required ten days, so it dies, as it were, "in his pocket").

Reaction to the bankruptcy law changes has been swift and fierce. Judge Frank Monroe of Austin, Texas, found himself compelled to reject the bankruptcy petition of Alfonso Sosa, a house painter who made $20,000 a year. He had filed for bankruptcy in the hopes of avoiding foreclosure on the $33,000 mobile home where he lived with his family. Judge Monroe noted that the new law is filled with traps for consumers, calling its provisions "inane" and "absurd." "Apparently," he observed wryly, "it is not the individual consumers of this country that make the donations to the members of Congress that allow them to be elected and reelected and reelected and reelected." [14]

Two weeks before the bill's passage, 104 bankruptcy law professors wrote to Congress complaining about the law. And the *New York Times* noted that critics of the bill "said the measure was a thinly disguised gift to banks and credit card companies, which, they contend, are largely responsible for the high rate of bankruptcies because they heavily promote credit cards and loans that often come with large and largely unseen fees for late payments. They said that the measure would impose new obstacles on many middle-income families seeking desperately needed protection from creditors and that it would take far longer for those families to start over after suffering serious illness, unemployment, and other calamities."[15]

Noting that the bankruptcy law revisions "provide for no distinction between those who get unlucky in Las Vegas and those who get cancer," Jonathan Alter of *Newsweek* charged that the revisions were "literally written by the credit-card industry."[16]

In an emotional piece tying the bankruptcy law changes to a hypothetical case, Alter said, "Let's say Peter Jennings was named Jeter Pennings and instead of making more than $7 million a year, he earns $70,000, still comfortably middle class. Pennings has lung cancer, and he understandably wants the best treatment available. But his insurance company won't cover experimental chemotherapy, so Pennings has an excruciating but familiar choice: he can charge the $25,000 chemo on his credit card or go without the cutting-edge treatment. If Pennings is like most people, he chooses to put his health first. With credit-card interest and late fees often totaling 100 percent a year, he's now so deep in the hole he'll never dig out. But under current law, he can file Chapter 7 and get on with what's left of his life." The piece went on to describe how Pennings would be unable to do so if, under the new law, he found himself in Chapter 13 bankruptcy.[17]

How did this law make it through Congress? How did it manage to attract enough Democratic votes in the Senate to avoid a filibuster?

The votes were bought and paid for. According to the *Washington Post,* the banking, credit card, and retail industries, "which have pushed for the legislation for more than seven years," gave more than $56 million in campaign contributions in the 2004 elections.[18] And, in the 2005–06 election cycle, while the bill was under consideration, the banking and finance industries gave $34 million more.[19]

The *New York Times* noted that "the main lobbying forces for the bill—a coalition that included Visa, MasterCard, the American Bankers Association, MBNA America, Capital One, Citicorp, the Ford Motor Credit Company, and the General Motors Acceptance Corporation—spent more than $40 million in political fund-raising efforts and many millions more on lobbying efforts since 1989. . . ."[20] All told, the industry has spent more than $150 million since 1990 in campaign contributions to elected officials and candidates in its effort to change the bankruptcy law.[21]

One key recipient of industry largesse, predictably, was the law's Senate sponsor, Senator Chuck Grassley (R-IA). Senator Grassley got the following campaign contributions from forces backing the bill:

FOLLOW THE MONEY:

CAMPAIGN CONTRIBUTIONS

TO SENATE SPONSOR CHUCK GRASSLEY[22]

- American Bankers Association PAC (BANKPAC) 1997 to 2005 — $18,000
- American Express Company PAC (AXPPAC) 1997 to 2003 — $15,000
- B of A Corp Federal PAC 1997 to 2006 — $38,500
- Citicorp Voluntary Political Fund Federal 1997 to 1998 — $ 4,000
- Citigroup Inc. PAC 1997 to 2004 — $11,499
- Mortgage Bankers Association PAC 1997 to 2006 — $ 8,000
- U.S. Bankcorp Political Participation Program 1998 to 2003 — $11,000
- Wells Fargo and Co Employee PAC 1997 to 2003 — $20,000

TOTAL TAKE:	$126,000

One expects the Republican Party to do the bidding of its upscale, corporate, special interest funders. But where was the Democratic Party on this bill? Except for a handful of Democrats, they put up no serious fight. Indeed, nineteen Senate Democrats voted for the bill, led by the new Democratic majority leader, Harry Reid (D-NV). These Democrats deserve special scorn. Despite being in the minority in the Senate, they could have killed the bill, as they killed many others, by filibustering and blocking the Republicans from getting the sixty votes they needed to close debate. Filibusters—once a rare tactic usually used only by Southern senators to kill civil rights legislation—have become a standard tactic in our political arena. If you don't have sixty votes, you don't bring a bill to the floor of the Senate. A majority is no longer necessary; you need sixty votes. But the GOP sponsors of the bankruptcy bill had no difficulty getting those sixty, because the financial industry had spread its money around to the Democrats who might have blocked the bill. The following table lists those Democratic senators who sold out and supported the bankruptcy bill, and how much they got in campaign contributions from commercial banks in the 2001–2006 period:

SENATE DEMOCRATS WHO SOLD OUT ON BANKRUPTCY PROTECTION . . . AND HOW MUCH THEY GOT[23]

2001–2006 Total

Arkansas	Blanche Lincoln	$145,507
	Mark Pryor	$59,396
Colorado:	Ken Salazar (new in 2006)	$9,500
Delaware	Joe Biden	$48,725
	Tom Carper	$298,018
Florida	Bill Nelson	$93,899
Hawaii	Daniel Inouye	$ 5,950
Indiana	Evan Bayh	$220,544
Louisiana	Mary Landrieu	$85,250
Michigan	Deborah Ann Stabenow	$150,096
Montana	Max Baucus	$166,727
Nebraska	Ben Nelson	$127,300
Nevada	Harry Reid	$109,896
New Mexico	Jeff Bingaman	$40,631
North Dakota	Kent Conrad	$114,500
South Dakota	Tim Johnson	$228,022
Vermont	Jim Jeffords (I) (not running)	$1,400
West Virginia	Robert Byrd	$66,450

Source: Profiles of individual senators on politicalmoneyline.com, listing the top twenty industries contributing to each candidate. (Some profiles did not include commercial banks among their top twenty industries, so we added these.)

Max Baucus of Montana is from a small state with no real opposition, but he still rakes in plenty of campaign money from major industries interested in his vote. In his most recent race, Baucus won with 63 percent of the vote. He's likely to win by even more next time. His reelection is a given, conceded by both parties. The donors didn't give him money because they were worried he wouldn't be reelected. They gave him the money to get his vote. And it worked!

Go down the list of Democrats. It's not hard to spot the more obvious examples of influence. Tom Carper from Delaware, for example, is not vulnerable—to put it mildly. He won last time with 56 percent of the vote and he will probably not face serious opposition next time. To give him almost half a million dollars in campaign contributions for an almost nonexistent campaign clearly smacks of buying access, if not influence. And, even if he needed the money for a campaign, Delaware is one of the smallest states in the nation. For half a million dollars, he could wallpaper the state with ads. But he won't. He'll use the extra money to pay for travel, staff, and other allowable perks of the job.

And that's an outrage!

ACTION AGENDA

The politicians have made a simple calculus: people who file for bankruptcy don't vote. Typically young families or individuals just starting out, they are burdened by student loans and the costs of a first home—and their turnout in elections is notoriously low. It's a truism in politics that the older you are, the more likely you are to go and vote.

So the congressmen and senators who voted to sabotage the bankruptcy law are expecting those who need its protections to stay home. They figure they will have a better chance of getting reelected if they pocket the campaign contributions and bet that those who are being hurt by the new law won't know about how they voted, or won't go out to vote themselves.

It's our job to confound them! Study the list, particularly of the Democrats who sold out. Send them a letter or an e-mail telling them you've heard about how they voted on bankruptcy and that they can't expect your vote in the next election. Get in touch with others who are similarly situated, and start a large movement. Don't let them get away with their vote! Stand up and fight. And let them know you're fighting.

Congress should

- let bankruptcy judges take the circumstances of the debt into account. If it was drugs or drinking, no relief. But if a petitioner can prove that medical bills or job loss caused his indebtedness, the judge should be allowed to use some discretion. Specifically, Congress should change the law so that a judge can forgive up to 80 percent of a debt (the current figure is only 20 percent) under Chapter 13 if the reason for defaulting on the debt was legitimate. Why should the rest of us bear this cost? Because we believe in second chances and forgiveness and mercy where it is deserved—and that's a worthwhile thing to spend our money on!

- Bankruptcy judges should be required to distinguish between the actual amount of the debt, the accumulated interest due, and the penalties imposed by the lender for late or missed payments. Not all debt is created equal. Judges should be able to forgive all penalties, and much of the interest, if the debtor had a good reason for his inability to pay.

- In repeated filings under Chapter 13, a judge should be able to forgive debt entirely if there is evidence that the debtor made a good faith effort to make payments in his previous period under Chapter 13. If a debtor can prove, for example, that he or she paid a certain percent of their income to extinguish the debt but just didn't make enough money to do it completely, a judge should be able to recognize the attempt and grant relief.

- Congress should turn its attention to regulating credit card companies and curbing their current practice of luring students and others with limited means into debt. Because some lending companies find that they make much more from bad debtors—with their accrued interest and penalties—creditors often deliberately seek out bad borrowers as targets for their credit card marketing.

- The absurd credit-counseling requirement should be abolished. This holier-than-thou attitude—coming from a Congress that bounced House Bank checks to high heaven a decade ago—has no place on the statute books. Only about 1 percent of those counseled enter debt management agreements; why the boondoggle for the counselors who collect fifty dollars a session?

And if our good legislators are so upset about bankruptcies, why don't they write some tougher legislation to crack down on corporate bankrupts like Enron, WorldCom, Global Crossing, and the like, who fleece tens of millions of dollars out of their employees' retirement savings?

Come on, Congress. Go for broke.

THE GREAT STUDENT LOAN ROBBERY

The Republican members of Congress claimed that theirs was a family-friendly agenda. But their hypocritical hiking of student loan interest rates belies their cruel pretensions of compassion—another reason young voters backed Democrats in 2006!

On February 8, 2006, President Bush signed what he called the "Deficit Reduction Act of 2005." In the accompanying press release, the president said that the bill was about "Improving Federal Student Loan Programs and Increasing Benefits to Students."[1] And yet these "improvements" actually *cut* student loan funding—by a net of $13 billion!—while socking student borrowers with huge interest rate increases which they must pay in order to get the education that the government keeps telling them they need to compete in today's global economy. When they graduate, these young debtors will feel the crunch of higher interest rates, which will hit young families just as they're trying to buy first homes, have children, and firm up their careers.

In fact, of the $40 billion in spending cut under the "Deficit Reduction Act" Bush signed, the largest share by far came from the cuts in student loans. This thoroughly iniquitous piece of legislation rolls back decades of federal policy bent on encouraging students to go to school and continue their education.

Here's what this horrible bill did:

PUNISHING STUDENTS AND FAMILIES:

The New Student Loan Legislation

- Raised interest rates on new student loans from 4.7 percent to 6.8 percent!

- Prohibited those currently holding student loans from refinancing or consolidating them to take advantage of future lower interest rates. They'll still have to pay 6.8 percent, even if rates drop well below that—making student borrowers the only class of debtors who cannot take advantage of lower rates!

- Raised interest rates on those who already hold student loans. For kids now in school, the rate will jump from 4.7 percent to 6.5 percent, and for those who have left and are struggling to repay the loans, it will go from 5.3 percent to 7.1 percent. Supporters of the rate hike point out that the student loan statute required an automatic increase when the 90-day Treasury interest rate increased. True, but the Congress did nothing to offer relief from this increase. And the ban on refinancing loans ensures that when interest rates drop, what went up won't come down!

- Raised interest rates for loans to parents of undergraduates, to finance their children's higher education, from 6.1 percent to 8.5 percent.

So much for family-friendly Republican policies! (And special thanks to Vice President Dick Cheney for casting the tie-breaking vote to pass this bill!)

In the opening days of the Democratic 110th Congress, the House passed a cut in student loan rates for Stafford Student Loans. Stafford loans are given to students based on need, are federally subsidized, and account for two-thirds of all student loans. The House bill would cut the interest rates as follows:

SAVING THE DAY: HOUSE CUTS STUDENT LOAN RATES[2]

July 1, 2007   Rate goes to 6.12%
July 1, 2008   Rate goes to 5.44%

July 1, 2009   Rate goes to 4.76%
July 1, 2010   Rate goes to 4.08%
July 1, 2011   Rate goes to 3.40%

However, the House bill offers relief only to those whose loans are subsidized based on their need. Each year, the proportion of such student loans has dropped. While two-thirds of all outstanding loans are subsidized and covered by the House bill, only half of the new loans made in 2006 were subsidized. The other half—those without subsidy or means testing—will face the draconian increase in interest voted by the Republicans last Congress. Since these families are largely still middle income and find the rapidly increasing college tuition hard to pay without borrowing, they deserve help from Congress as well.

The House bill has yet to pass the Senate or be signed by the president. Ominously, President Bush has already announced his opposition to the bill, so it's unclear what the outcome will be. Was the House action just a piece of one-house propaganda, or a real and badly needed reform? Stay tuned (and see the action agenda at the end of this chapter for what you can do!).

In 2006, more than $60 billion in federal student loans went to about eight million students. Forty-four percent of all college students take out student loans.[3] Eighty-four percent of black students and two-thirds of Latinos graduate with debt.[4] The average student borrower now leaves college with a debt of $27,600-three and a half times the average ten years ago. These increases of two points in interest will sock these average students with an extra $500 annually—just as they're trying to start a family, begin a new job, and buy a first home. These young people will find it harder to finance their educations or to repay the loans after they leave school—because of the "improvements" to the student loan program passed by this Congress.

These interest rate increases, of course, come on top of average increases in private college tuition and fees of 37 percent, and a rise in public college tuition of 54 percent over the past decade. That's why two-thirds of all college graduates incur debt to get their diplomas.[5]

To put the best possible face on this student gouging bill, the then House majority leader, John Boehner (R-OH), said that the law offers "significant new benefits" to students, "from lower loan fees to higher loan limits to a simplified financial aid application process."[6] Some benefit! The bill cut spending for student loans by $20 billion through higher interest rates—a gift that keeps on giving—and claimed to offset that cut by about $8 billion in new benefits. Comments like that are part of the reason why Boehner was demoted to minority leader!

These "benefits" include a reduction in some fees students have to pay on their loans, from the current 4 percent to 1 percent, by the year 2010. That's a 3 percent, one-time cut, at a time when Congress has raised the interest rates by 2 percent annually. Some "significant new benefit"!

(As the *Harvard Crimson* has reported, the bill also permitted "students majoring in math, science, and certain languages, as well as undergraduates from 'rigorous' high schools" to get grants of up to $4,000. The bill also raised federal loan limits by about $900 for college freshmen and $1,000 for sophomores—at a time when college costs are rising by an average of over $1,500 annually.[7])

Having increased at an annual rate of 5.3 percent for the past decade, college costs now average $29,000 a year.[8] One who recognizes the magnitude of this problem is Richard A. Davies, senior managing director of Retirement and College Savings Plans for AllianceBernstein Investments. "College financing is a challenging issue for most American families," Davies says, "and the fact that college costs are climbing much faster than the rate of inflation does not make it likely that the problem will go away anytime soon."[9]

Students are now graduating with huge and staggering debt levels that have a sharp and immediate impact on their ability to start their lives. According to the AllianceBernstein study, one-third of indebted college graduates say they have sold possessions to make ends meet while almost half say they are living "paycheck to paycheck." Davies notes that "large amounts of college debt put graduates in a hole that can take years, even decades, from which to emerge."[10]

STRUGGLING STUDENTS: HOW STUDENT LOAN RECIPIENTS ARE MAKING OUT [11]

- 62 percent of college graduates aged 21 to 35 still have debt
- 39 percent said it would take ten years or more to be debt free
- 43 percent used credit cards to pay tuition and one-third are still in debt
- 53 percent say higher rates will have a "major impact" on their ability to pay
- 29 percent miss payments occasionally or regularly
- 44 percent say the debt made them delay buying a home
- 28 percent say the debt made them delay having children
- One-third say the debt made them move in with their parents
- One-quarter had to postpone or cancel graduate school attendance

Source: "The College Debt Crunch," AllianceBernstein Investments, May 26, 2006.

Why was Congress so anxious to sock it to the students and their parents? According to Eileen Ambrose, a columnist for the *Baltimore Sun,* it had a lot to do with disgruntled bankers. She explains that the "federal government promises lenders a minimum return on student loans." When families consolidate their loans and take advantage of lower rates, the government "spends billions subsidizing" the lenders because of the lower rates they get. As Ambrose notes, this "didn't sit well with some members of Congress." [12]

Fine. But, instead of curtailing the subsidy, the government simply mandated that students and parents pay higher interest rates. They kept the subsidy intact!

Then Congress made sure that loan consolidation would stop, so lenders could keep earning higher interest rates on student loans—much higher than what they make from mortgages and business loans that borrowers can refinance. "Large lenders also lobbied hard for changes that would make consolidating less appealing to borrowers because those loans are the least profitable in the federal program," according to Mark Kantrowitz, who publishes financial aid information.[13]

For Congress, it came down to a decision: to help bankers or students. They chose to continue their aid to bankers and raise interest rates for students. Why aren't we surprised? Student loans are, after all, big business. As of August 2006, according to *Bloomberg News*, federal and private loans "constitute an $86 billion business, accounting for 39 percent of the $222 billion spent on higher education."[14]

But why do private-sector lenders provide student loans at all? As *60 Minutes* noted in 2006, "There is a less costly way to make student loans, called 'the Direct Loan Program,' run by the Department of Education. . . . Studies by three different government agencies say it costs taxpayers about five times less per student loan" to have the program run by the government directly.[15]

Bill Clinton tried to make all student loans flow directly from the Department of Education, passing on banker profits to students in the form of lower interest rates. But the special interests—and their congressional allies—turned him down. Adding to the injury, they are now raising the interest rates that private lenders charge.

So why do students borrow from private lenders when the terms offered by the government are more generous?

The answer: kickbacks.

As *60 Minutes* describes it, "Most universities guide their students to private lenders. . . One reason is because they offer incentives to some of the schools." Under the "school-as-lender" program, private lenders pay universities money to help administer their loans; in return, universities steer students to the private borrowers[16]

Particularly notorious for exploiting this loophole was the organization known as Sallie Mae, originally a public corporation Congress established to create a secondary market to buy up student loans from banks and other lenders. In 1997, Congress authorized the privatization of Sallie Mae, a process that was finished in 2004. As a private entity, Sallie Mae manages $123 billion in student loans with ten thousand employees. Its stock price has risen 2,000 percent since 1995.[17] One of the reasons Sallie Mae gets so much business is that it generously provides "incentives" to colleges to push their student loans.

Trying to level the playing field, Congressman Thomas Petri (R-WI) sponsored legislation to permit the Direct Loan Program, run by the federal Department of Education, to offer incentives to colleges the way the private lenders do. "All we were asking is that the lower-cost program be allowed to compete on a more equal basis," Representative Petri explains. "And we think the result will be that we'll save money for the taxpayer. And it will provide exactly the same benefits and a few more to students."[18] Sallie Mae lobbied hard and killed the Petri amendment.

One of the reasons Sallie Mae and other private student loan lenders make out so well is that Congress guarantees the lenders ninety-seven cents on the dollar if a borrower defaults. Beyond the 97 percent of the principal, and the compounded interest, it can also seek to collect the loan from the borrower as well—and if it is successful, it gets to keep up to one-fourth of whatever it recovers.

Here's how it works: let's say you borrow $12,000 from Sallie Mae for college, and then you default. Sallie Mae hits up Uncle Sam for 97 percent of the $12,000, or $11,640, plus compounded interest. Let's say that after years of dunning and lawsuits, Sallie Mae manages to force you to pay back the $12,000. Sallie Mae keeps $3,000 of the money you paid back. So it gets $3,000 + $11,640 + compounded interest. Pretty neat for them—and an outrage for the rest of us!

As *60 Minutes* reports, in 2005 "nearly a fifth of [Sallie Mae] revenues came from its collection business."[19] Elizabeth Warren, professor of bankruptcy law at Harvard Law School, put it archly: "Sallie Mae makes money if you pay back on time. And Sallie Mae makes money if you don't pay back on time."[20]

At the heart of Sallie Mae's success is its lobbying skill. Since 2002, the company and its employees have donated $2.7 million to candidates for Congress.[21] House Minority Leader John Boehner (R-OH) is one of Sallie Mae's guardian angels. He received more than $200,000 in campaign donations from the private student loan industry in the 2003–04 election cycle.[22]

The banking industry's lobbying to pass the bill raising student loan interest rates was to be expected. But the shocker is that college financial aid officers joined the chorus—on the wrong side. Why? They have been getting kickbacks. According to the *Washington Post,* "many school financial aid offices operate symbiotically with lending institutions, accepting gifts of equipment and other perks in exchange for directing students their way." The lenders wanted payback from the financial aid counselors: help to jack up the interest rates students have to pay![23]

Of course, this all raises a question: why are private lenders involved in the student loan business, anyway? As the *Washington Post* noted, "study after study has shown that direct loans are cheaper for taxpayers than loans that flow through middlemen. But because of the incentives that those middlemen offer colleges, not as many students wind up using the direct loans."[24]

All this misguided federal policy is the direct result of campaign contributions from the banking and finance industry—one of the country's most generous donors to needy politicians, if not to students. The industry gave a total of $52 million to politicians in the 2005–06 election cycle alone. Massive funds went to both parties: $32,317,230 to Republicans, and $18,975,614 to Democrats.[25]

CONTRIBUTIONS TO POLITICIANS BY INDUSTRIES EAGER TO SEE STUDENT INTEREST RATES RAISED

Donations to Congressional Candidates from the Finance and Banking Industry, 2005–2006[26]

Finance, Insurance	DEMs	REPs	OTHER
1. accident and health insurance	$2,067,911	$3,804,299	$4,500
2. accounting/auditing/ bookkeeping services	$1,639,106	$3,769,811	$15,110
3. actuaries	$55,076	$33,500	$0
4. banks	$3,795,156	$7,254,157	$13,000
5. boards of trade	$11,000	$6,000	$0
6. check cashing service	$14,500	$43,650	$0
7. commodity brokers	$57,100	$78,000	$0
8. commodity exchanges	$764,000	$864,800	$1,000
9. credit card and other credit plans	$396,899	$565,124	$000

10. credit reporting agencies	$210,553	$461,900	$0
11. credit unions	$1,281,549	$1,537,952	$27,950
12. federal/federally sponsored credit agencies	$87,700	$777,350	$12,500
13. financial planning consultants	$77,750	$125,500	$0
14. fire, marine, and casualty insurance	$100,819	$286,750	$0
15. holding companies (bank)	$0	$16,000	$0
16. holding companies (nonbank)	$14,000	$14,000	$0
17. importers	$1,000	$0	$0
18. insurance agents, brokers, and service	$397,300	$1,211,975	$0
19. insurance carriers	$908,147	$1,942,218	$3,500
20. investment bankers	$265,100	$564,450	$0
21. investment management	$671,603	$912,155	$1,000
22. investors	$19,200	$68,398	$0
23. life insurance	$1,942,737	$2,666,557	$9,500
24. loan brokers	$161,225	$227,935	$11,000
25. miscellaneous finance	$83,000	$157,350	$0
26. mutual funds	$74,500	$85,500	$0
27. pension health and welfare funds	$6,500	$65,100	$0
28. personal credit institutions	$571,297	$1,020,774	$2,000
29. real estate investment trusts	$365,000	$539,610	$0
30. real estate loans	$559,197	$653,784	$15,000
31. savings and loan associations	$188,080	$270,011	$0
32. stock and bond brokers	$905,109	$1,378,942	$0
33. stock exchanges	$158,500	$152,412	$0
34. tax return preparation services	$74,000	$78,500	$2,000
35. title insurance	$89,000	$175,766	$1,500
36. venture capital	$262,000	$507,000	$0
Total	$18,975,614	$32,317,230	$120,560

Derived from latest data key-punched into FEC databases as of 01/1/2007.

Source: PoliticalMoneyLine, from "Finance & Insurance Industry Totals," Congressional Quarterly, http://www.fecinfo.com/cgiwin/x_ee.exe?DoFn=06F.

The breadth and scope of the banking and finance industries' mobilization to lobby Congress is appalling—to everyone except congressmen favored by those industries, who saw it as yet another opportunity to enrich their campaign coffers.

The fact is that almost every facet of the industry has turned itself into a cash cow for Congress. But not for just *any* member of Congress. The industry aimed its money wisely, concentrating on the House and Senate banking and finance committees, which consider the legislation that concerns the industry the most—such as the student loan "reforms."

Indeed, the major reason members of Congress choose to sit on the banking committees is precisely to get those lucrative campaign contributions. Those who want to appease the farmers in their districts naturally select the Agriculture Committee. The members who want to court labor usually end up on the Education and the Workforce Committee. Those who want defense dollars for their districts turn to the Armed Services Committee. And those who want to exercise international influence—and possibly cement qualifications to run for president—choose the Foreign Affairs Committees.

But there are few constituent interests at stake in the banking committees' deliberations, unless one counts the harm that comes to the average voter from special interest giveaways to banks. In Washington, those who join these committees are notorious for using them to rake in contributions from the special interests who are especially concerned about legislation before the committees.

Senate Banking, Housing, and Urban Affairs Committee members received almost $3 million from commercial banks in the 2005–06 election cycle. Here's the list of the favored few in the Senate who revel in the largesse of the finance industry—and, as a result, express their thanks through passing special interest legislation like the student loan "reforms."

WHO GOT THE MONEY?

Commercial Bank Donations to Senators on the Banking Committee [27]

Charles E. Schumer (D-NY)	$547,599
Richard Shelby (R-AL)	$372,073
Elizabeth Dole (R-NC)	$312,530
Tom Carper (D-DE)	$310,936
Rick Santorum (R-PA) (since defeated)	$321,850
Tim Johnson (D-SD)	$241,372
Chris Dodd (D-CT)	$226,900
Evan Bayh (D-IN)	$227,144
Jim Bunning (R-KY)	$165,249
Mel Martinez (R-FL)	$262,052
John Sununu (R-NH)	$180,650
Debbie Ann Stabenow (D-MI)	$157,096
Mike Crapo (R-ID)	$141,175
Jack Reed (D-RI)	$144,728
Wayne Allard (R-CO)	$137,204
Robert Bennett (R-UT)	$126,124
Mike Enzi (R-WY)	$94,950
Chuck Hagel (R-NE)	$88,550
Paul Sarbanes (D-MD) (since retired)	$11,750

Source: Center for Responsive Politics at www.opensecrets.org, specifically "Member Money: Top Committee-Related Industries for Committee Members—Banking, Housing, And Urban Affairs, 109th Congress, 2006 Cycle Data," at http://www.opensecrets.org/cmteprofiles/profiles.asp?cycle=2006& CmteID=S06&Cmte=SBAN&CongNo=109&Chamber=S

Similarly endowed, members of the House Financial Services Committee got a total of $2.8 million in the 2005–06 election cycle from commercial banks alone, and a total of $12.2 million from the general finance, insurance, and real estate industry.[28]

ACTION AGENDA

Write your congressman to protest the higher student loan rates. Make them feel the heat! The Congress that prides itself on cutting taxes and helping "the average" American—and the administration that says it accords a top priority to helping education—should not be so blithely able to raise the cost of a college and postgraduate education. So far, the borrowers have barely whimpered. There needs to be a roar!

Particularly, lobby the Senate to pass the reduction in student loan interest rates passed by the House, and write President Bush to urge him to sign the legislation. Every former or current student, and every parent with a student in school, should write such a letter. It will save you or your child thousands of dollars!

Write your House member—particularly if he is a Democrat—and demand that the House extend its interest relief to all student borrowers, whether subsidized or not.

When you take out a student loan, be sure you go directly to the government and borrow from the federal Department of Education. Don't let your friendly college financial aid officer steer you to a higher cost private lender.

Finally, demand that Congress

- permit student borrowers to consolidate their loans and refinance them at lower interest rates whenever they can, just as other borrowers do. Lenders should not be allowed to discriminate against student borrowers;

- end private student lending by channeling all student loans through the Department of Education. We know all the arguments about government inefficiency, but this is one example where the public sector can make loans cheaper than private business can, because Uncle Sam doesn't make a profit. The government guarantees 97 percent of the loan anyway, so let the government make the loan in the first place!

- ban lenders' payments of "administrative fees" to college admission and financial aid offices. These thinly disguised bribes, designed to get the officers to steer students to higher-cost lenders and away from lower-cost government loans, must end; and

- with interest rates going up, cushion the increases for student loans. Clearly, a college education and graduate training distinguishes our labor force from those of other countries, and it permits lifelong upward mobility. Today's higher interest rates are a disincentive to higher education—which is the last thing our country needs. We're not asking for a subsidy—just something that feels less like punishment!

KATRINA INSURANCE SCAMS:
"What Do You Mean I'm Not Covered?"

On August 29, 2005, Hurricane Katrina swept into New Orleans and the Gulf Coast. Katrina smashed the city's levees, flooded its streets, and demolished hundreds of thousands of homes in a swath from New Orleans to Pascagoula and Biloxi.

As soon as the winds and flooding died down, adjusters and agents for the big insurance companies also swept through the region, frequently destroying what little hope the battered families had left after the storm. Insurers like Nationwide ("We're on your side"), State Farm ("Like a good neighbor"), and Allstate ("You're in good hands"), delivered the news to the bewildered homeowners: they were not insured for water damage from Katrina. Having neglected to read the fine print on their homeowner's policies—and often relying on brokers' misrepresentations—they found, to their horror, that the insurance companies would pay only for damage brought about by wind, not for water damage associated with the Katrina storm surge. In fact, if a home had any water—even just a millimeter—after Katrina, the insurance companies were even trying to weasel out of paying for wind damage at all.

Just before this book went to print, State Farm finally agreed to settle the claims against it by Mississippi homeowners for a total that Mississippi attorney general Jim Hood says may reach $500 million. Included will be $80 million in payouts to six hundred especially hard hit Mississippi homeowners and, at least $50 million to those represented by famed anti-tobacco attorney Dick Scruggs, who sued State Farm and forced the settlement.[1]

But State Farm is still thumbing its nose at many of its Louisiana policyholders who were victimized by the storm. And as we go to press, Nationwide and Allstate have still not settled. We hope they see the light. Maybe this book will help!

Legally, the insurance companies had a case. They had hornswoggled hundreds of thousands of families that were at risk from hurricanes into buying, in effect, insurance against half the storm: the dry part. Their argument recalls that of the lender in Shakespeare's *Merchant of Venice*. When a borrower cannot repay his debt on time, the lender demands the heart as his "pound of flesh." He loses his argument in court, since the loan agreement specifies only that a pound of flesh is to be taken, but says nothing about the blood flow that will accompany it.

Hurricanes bring with them wind and rain and a surging tide. Their combined effect is horrific. By only covering the flesh, not the blood, the insurance companies drew a distinction that is, by definition, impossible to apply to a hurricane.

So literal were the insurance companies in their attempt to delineate between wind and water damage that, as the *New York Times* reported, "Many homeowners have complained that they received less than they expected for wind damage because they were told by their insurers that they were entitled to receive payments only for damage to roofs and other parts of their house above the highest water stains on their walls."[2]

But testimony by meteorologist Rocco Calaci indicated that "hurricane-force winds battered the coast for hours before water rushed over the land, knocking down houses and trees. "High winds," he said, "continued for several more hours after the water had receded. Several people who rode out the storm in their homes testified earlier that, in their neighborhoods near the Alabama border, the water rose slowly and gradually to more than five feet in their homes. . . . All along the Mississippi coast, residents say, their homes suffered heavy wind damage long before the water began to rise."[3]

By applying the high-watermark standard to exclude coverage of any damage below the ultimate waterline, the courts may be precluding the justifiable claims of homeowners whose buildings were damaged by wind before they flooded with water.

Insurance companies counter that homeowners can buy flood insurance from the federal government. And, indeed, more than $15 billion in federal flood insurance payments have gone out in the wake of Katrina.[4] Wind damage claims are expected to cost private insurance companies up to $30 billion when the costs of Katrina are totaled up.[5]

But insurance agents, anxious to sell their company's policies, failed to warn thousands of ordinary homeowners that their policies did not cover water damage. They typically chose not to undermine their sales pitch by advising consumers to buy federal flood insurance as well.

Attorney Richard Scruggs, who brought big tobacco to heel in the attorney general's lawsuits, represents many of these disgruntled homeowners. He told PBS: "The fact of the matter . . . is that the commission that the insurance agent makes on flood insurance is only about forty bucks. It's not worth the paperwork. That's why they don't promote it."[6]

Typical of the conundrum homeowners faced after Katrina is the situation of Scruggs's clients Paul and Julie Leonard of Pascagoula, Mississippi. Their home sustained $130,000 of damage in the storm, but they could only trace $3,000 of this sum to wind alone.[7] Why hadn't the couple purchased federal flood insurance to supplement their Nationwide policy? Because they had been misled by an unscrupulous insurance agent who was out for a sale.

The *New York Times* reported that "Mr. Leonard, a police lieutenant, testified that the agent had told him that he did not need to buy flood insurance from the federal government. Mr. Scruggs argued that Mr. Leonard . . . had reasonably concluded that his home insurance policy provided everything he needed to protect him in a hurricane."[8] The *Times* noted that Scruggs said that "Nationwide and other insurers led customers to believe that any hurricane damage—whether from wind or water—would be covered. 'They thought they had all [the coverage] they needed.'"[9]

Even the Leonards' own insurance agent, himself a Gulf resident, had not bought flood insurance. In fact, only 180 of the 1,200 Nationwide customers to whom he sold policies had purchased flood insurance. In all, fewer than 20 percent of the homeowners along the Mississippi coast had done so. The insurance companies had, obviously, widely and successfully peddled the notion that their policies were adequate for all eventualities.[10]

Scruggs sued on behalf of the Leonards in Gulfport, Mississippi, in what was widely regarded as a test case. United States District Court judge L. T. Senter Jr. sided with the insurance giants against the Gulf Coast families. The *Times* reported that while "Judge Senter said that . . . he believed that the [Leonards' insurance] agent had advised against buying flood insurance, he did not agree that the agent had, in effect, verbally expanded the coverage of the home insurance policy to include flood damage."[11]

The moment of truth came when Paul Leonard testified that he had asked his insurance agent if he needed flood insurance for his home, which is two blocks north of the water. The police officer said that the agent told him he did "not need that s – – t." The judge dismissed the agent's statement as "advisory in nature," and found that there was "insufficient evidence" that the agent had misled Leonard."[12]

In his opinion, Judge Senter said, "The plaintiffs are bound by the express terms of the policy, even if their interpretation of the policy was incorrect and even if the inferences they drew from the conversations they had with [insurance agent Jay] Fletcher were erroneous."[13]

Of course, we're not all experts in insurance law. We rely on the companies' agents to explain the coverage and tailor it to our needs. In this case, clearly, the Nationwide agent had made it clear that his policy rendered flood insurance unnecessary. Judge Senter's narrow, legalistic ruling profits insurance companies at the expense of his neighbors on the Coast. It is absurd to think that any homeowner in his right mind would buy a costly private policy that did not cover flooding, and then turn his nose up at cheaper federal flood coverage—unless he had been misled by an insurance agent into believing that the private policy was sufficient because it covered flooding.

The insurance companies must have known that their agents were predicating their sales on overstating the coverage and minimizing the need for federal insurance to supplement it. For Judge Senter to permit such shenanigans, without calling the insurance companies to account, is an outrage.

Dr. Munson Hinman, a chiropractor and Katrina victim, had a similar experience with a Nationwide agent. In a separate litigation, he testified that his broker "talked him out of buying [flood] coverage six weeks before the hurricane struck. [He said] in a roundabout way . . . it wasn't necessary. He was emphatic . . . that it wasn't necessary."[14]

Judge Senter did not, however, permit Nationwide, Allstate, and State Farm to abuse their customers as much as the companies wanted him to. The insurers had sought a ruling that would exempt them from making *any* payment—even for wind damage—if water damage was present as well. Scruggs said that by rejecting the insurance companies' contention, this aspect of the court ruling was "going to open the door for recovery for thousands of Mississippi homeowners."[15]

Homeowners in Louisiana were luckier in their dealings with both big insurance and the federal courts—if not with Mother Nature. On November 29, 2006, federal judge Stanwood R. Duval Jr. of the Federal District Court in New Orleans ruled that insurers had to pay Louisiana hurricane victims because the New Orleans' flooding was caused by a break in the levees and, hence, human error, while the Mississippi damage was due to natural causes. In his opinion, Duvall pointed out that "language in the insurance policies on flood coverage was ambiguous because it did not 'clearly exclude man-made' flood disasters." He excepted State Farm and the Hartford Insurance Companies, because these two companies were prescient enough to exclude flood damage whatever the cause.[16]

The final decision on flood coverage for Katrina, both in Mississippi and Louisiana, will take years to complete. The higher federal courts will have to reconcile any differences between the Mississippi and Louisiana Federal Court rulings, and then trial lawyers will litigate the individual cases to determine whether the damage was caused by wind or water. In the meantime, the policyholders get nothing . . . And that's an outrage!

The lesson? Don't trust insurance companies!

TOBACCO COMPANIES:
More Addicts, More Profits

**NICOTINE LEVELS INCREASED BY 10 PERCENT
FROM 1998 TO 2004**

About half a million people die each year because of illnesses directly caused by tobacco. And now, the tobacco companies are trying to make it even easier for people to become addicted to their poisonous cigarettes— and to make it harder to quit. More addicts, more profits. That's what it's all about. So what if it causes more deaths?

In 1998, the tobacco companies were publicly brought to their knees. A landmark settlement with the states' attorneys general required them to pay more than $250 billion to cover massive health care costs from illnesses caused by tobacco. That's a lot of money, even for companies like Philip Morris and RJ Reynolds.

The industry was also forced to discontinue marketing and promotions aimed at kids, set up a foundation to monitor teen smoking, and develop a national advertising campaign about the dangers of smoking. The results have been extraordinary. The number of adults who smoke has dropped from 24.7 percent in 1997 to 20.9 percent today.[1]

The publicity about how big tobacco deliberately created a product that would addict its users to ensure an ongoing market for years to come did not endear the tobacco executives to the American public. And the disclosure that they knew all about the deadly illnesses caused by tobacco and intentionally concealed that information—and even denied it—made the same executives social pariahs.

You'd think they might have learned something from that experience. They didn't.

A recent study indicates that since the 1998 settlement, tobacco companies have been deliberately making it easier to hook people on cigarettes by quietly increasing the nicotine level that smokers can inhale in all brands— including those labeled "light"—by an average of 10 percent.[2]

The tobacco companies are run by truly evil people.

According to the National Institute on Drug Abuse, "Most smokers use tobacco regularly because they are addicted to nicotine."[3] Nicotine increases the release of dopamine in the brain pathways, creating an immediate sensation of pleasure and an adrenaline rush that is quickly diminished. In order to maintain that feeling of pleasure, smokers addicted to tobacco need to smoke more and more cigarettes. The more nicotine there is in each cigarette, the harder it is to quit smoking.

MANIPULATING NICOTINE
TO CAUSE CANCER[4]

- From 1998 to 2004, ninety-two cigarette brands raised their nicotine levels

- Twelve decreased it

- Twelve stayed the same

Source: Lois Keithly, PhD. et al, "Changes in Nicotine Yields, 1998–2004," Tobacco Control Program, Massachusetts Department of Public Health

---

The Massachusetts Department of Public Health study "Change in Nicotine Yields, 1998–2004," which tracked the nicotine levels in 116 brands of cigarettes, marks the first time that nicotine levels have been disclosed since the tobacco settlement. Massachusetts law requires the testing of all cigarette brands to determine the nicotine content.[5]

The worst offender was Newport filtered cigarettes, which registered the highest nicotine levels; Marlboro, Camel, Basic, and Doral were all close behind. Teen smokers were most likely to smoke the higher-nicotine-rated brands. According to the report, more than two-thirds of high school seniors choose Marlboro as their favorite cigarette brand, followed by Newport and Camel.[6]

Ironically, the tobacco settlement required tobacco companies to contribute more than a billion dollars to a foundation that would, among other things, track and monitor teen smoking and identify the reasons for both increases in smoking by young people and for failure to decrease their tobacco use. The companies were also required to pay for a long-term national advertising and public education campaign to educate the public about the cause and prevention of diseases associated with tobacco use.

But no one knew what the companies were really up to. They were not routinely required to disclose nicotine levels, and they definitely did not do so voluntarily. Contrary to popular impression, the federal government does not regulate tobacco companies. The companies report to no one but their shareholders.

Thus, while tobacco companies were compelled to spend a fortune to advertise about the dangers of tobacco to young people, they were also secretly placing products on the market that were intended to addict those same kids. It's easy to figure out why: because, unless the tobacco companies can hook more young people and retain their older customers, they'll be out of business when their older customers—who have been smoking for years—die.

And, of course, unless the tobacco companies make cigarettes more addictive, people will quit in greater numbers. More than fifty million Americans have quit smoking. Big tobacco is making cigarettes more habit-forming for exactly this reason: to ensure that this number does not grow.

This deception by big tobacco is not new. In August 2006, United States District Court Judge Gladys Kessler ruled that the tobacco companies had conspired for decades to cover up the dangers of smoking. Kessler found that the tobacco companies "suppressed research, they destroyed documents, they manipulated the use of nicotine so as to increase and perpetuate addiction . . . and they abused the legal system in order to achieve their goal—to make money with little if any regard for individual illness and suffering, soaring health costs, or the integrity of the legal system."[7]

Kessler also held that the companies make their profits by "selling a highly addictive product which causes diseases that lead to a staggering number of deaths per year, an immeasurable amount of human suffering and economic loss and a profound burden to our national health care system."[8]

She pointedly accused the tobacco companies of deceiving the public by claiming that nicotine was not addictive when they knew it certainly was. She said that they had concealed their own research that proved the addictive nature of nicotine. Kessler castigated the companies for deliberately making addictive products: "Defendants purposefully designed and sold products that delivered a pharmacologically effective dose of nicotine in order to create and sustain nicotine addiction in smokers."[9]

Judge Kessler also blamed the industry's marketing programs for leading kids to smoke, and prohibited them from labeling any cigarette brand as "light" or "low-tar" or any other term that falsely implies that such a brand is less dangerous. She rightly concluded that the big tobacco companies had "marketed and sold their lethal product with zeal, with deception, with a single-minded focus on their financial success, and without regard for the human tragedy or social costs that success exacted."[10]

They're still at it. And that's an outrage.

ACTION AGENDA

In July 2005, President Clinton and his FDA declared that nicotine was an addictive drug and that cigarettes were a drug-delivery device. Just as heroin is addictive and syringes are devices for delivering the drug, the FDA ruled that cigarettes were essentially tools of addiction. Based on this finding, it claimed the right to regulate tobacco.

Unfortunately, the Republicans blocked the ruling, and Congress refused to ratify the FDA's assumption of jurisdiction; without this congressional approval, the federal courts ruled that the FDA did not have the power to act.

Still, the FDA's discovery of how tobacco companies manipulated the nicotine content of their cigarettes was a seminal moment. For these companies, making cigarettes involved more than just harvesting tobacco leaves, chopping them up, and rolling paper around them. The FDA found that they had deliberately added chemicals to the cigarette to make it more addictive—hence its finding and assumption of jurisdiction.

Congress should look at the new findings of the Massachusetts agency that discovered that tobacco is continuing to pull the same old tricks in marketing its products to children. The House and Senate need to reconsider the legislation that stopped the FDA from taking jurisdiction. The lives of so many Americans, particularly young people, depend on it.

A TALE OF TWO CROOKS:
Gerhard Schröder and Jacques Chirac

When American journalists speak of "world public opinion," they usually mean Western European opinion or, more precisely, German, French, and British opinion.

But two of these countries have been led through most of the 1990s, and the early years of this decade, by men who should be in jail—not leading their nations. Germany's former chancellor, Gerhard Schröder and France's president, Jacques Chirac, have both been involved in dealings so shady that they unquestionably should disqualify them for the high positions they hold or once held.

As mayor of Paris in the 1980s, Jacques Chirac routinely accepted bribes from a wide variety of business interests who wanted special favors from his administration. One such businessman who personally paid Chirac a huge cash bribe, Jean-Claude Mery, revealed as much in a deathbed confession.[1] But since Chirac was then serving as president of France, there could be no thought of prosecution until he left office. Indeed, French libel laws are so strict in protecting public officials like Chirac that mention of this scandal is all but prohibited in the French media.

Schröder's scandal arose in the weeks after he left office. It involved his approval of a controversial gas pipeline deal with Russia—and his acceptance of a top-level job, within months of his departure from the chancellor's office, with the government-controlled pipeline company.

So what? Why should Americans care if the French and Germans want to elect crooks to run their countries?

We should care because the fact that they have so blatantly committed offenses that would likely land them in jail in the United States—and have completely gotten away with them—speaks volumes about the differences between American and continental European ethical standards in public office.

American scandals have always both amused and shocked European observers. Few understood what was so bad about President Nixon getting the CIA to cover up for the burglars his men (probably with his knowledge) sent to wiretap the Democratic National Committee. Even fewer grasped why we were so offended when another president lied under oath in a deposition in a federal lawsuit about his relations with a certain very young lady. By contrast, no French president since the start of the Fifth Republic, or German Chancellor since World War II, has ever been hounded out of office—or even seriously injured by scandal.

As the need for unified, coordinated global action on issues like terrorism, global climate change, energy security, drugs, nuclear proliferation, and immigration becomes ever more manifest, we have to know the leaders with whom we are dealing. While we expect corruption in Arab or third world countries, it comes as a shock when it appears at the very top of our closest allies' governments.

But their ways are not our ways, and their thoughts are not our thoughts. We need to realize how much more lenient and forgiving the French and German people, and their media, are. When it comes to official corruption, they're more likely to wink at it than to prosecute it. The vigorous policing of public conduct, orchestrated here by zealous United States attorneys and the public corruption office of the Department of Justice, is simply not a feature of continental European political life, as the stories of Chirac and Schröder make abundantly clear.

Let's examine Schröder first.

Schröder Sells Out Freedom for Cash

In September 2005, ten days before the German election, Chancellor Gerhard Schröder announced his government's approval of a deal with Russian president Vladimir Putin to help the government-controlled Russian gas company, Gazprom, build a $4.7 billion pipeline under the Baltic Sea to ship fuel from Russia to Germany. Schröder also approved a $1.2 billion line of credit for the venture.[2]

Ten days after the announcement, Schröder was defeated for reelection by the Christian Democratic Party, led by Germany's new chancellor, Angela Merkel. Then, in December 2005, a bare three months after he left office, Schröder signed on as a board member of the North European Gas Pipeline (NEGP) Company, the pipeline company whose venture he had just approved. He has since been elected its chairman—collecting a $310,000 salary.[3]

Gazprom, the Russian company, is the largest natural gas supplier in the world. The Russian government totally controls the company under director Alexandre Medevev, who is Putin's former chief of staff.

What makes Schröder's conduct disgusting, as well as reprehensible, is that the pipeline is part of an elaborate effort by Putin to enhance Russian power over the democratic, independent governments in Poland, Belarus, and Ukraine, which refuse to toe Putin's line.

The current pipeline that ships gas to Western Europe flows through Poland, Belarus, and Ukraine. Putin is determined to use that gas as a political weapon in his long-running battle with Poland and Ukraine.

Ukraine incurred Putin's wrath in 2004, when anti-Communist leader Viktor Yushchenko led his forces into the streets of Kiev for a two-month long demonstration to stop the Russian puppet government that ruled the country from stealing the election. Yushchenko—a client of ours—had won about 60 percent of the vote in the November 2004 election, but the government tried to pretend that Russia's favorite candidate, Viktor Yanukovitch, had been elected. Hundreds of thousands of people demonstrated for two months and forced a new election, which Yushchenko won handily.

Before the attempt at ballot fraud, the Ukraine KGB had tried to kill

Yushchenko by poisoning his soup with dioxin. It almost killed him and, where the poison sweated out of his body, left him with huge and permanent disfiguring scars on his face.

Putin was outraged when the new election was called, and even visited Ukraine to campaign for Yanukovitch. Thwarted by the voters of Ukraine, he vowed revenge.

Poland offended the would-be Russian czar in early 2005, when it ousted the former communists from power in Warsaw and elected dedicated democracy advocates, Lech Kaczyński as president, and his twin brother, Jaroslaw Kaczyński, as premier.

To punish Yushchenko on the eve of his country's parliamentary elections, and send a message to Poland, on January 1, 2006, Putin announced that he was cutting back gas deliveries to Ukraine. To do so, he ordered Gazprom to reduce the flow of gas in its pipeline by the amount Ukraine would normally have drawn off for its own use. The rest of the gas is usually sent on to Germany for shipment throughout Western Europe.

But Yushchenko had another move up his sleeve. He ordered the Ukrainian gas company to draw its usual compliment of gas from the pipeline, reducing the amount it shipped on to Germany—effectively cutting Western European supplies instead of his own. The European Union went apoplectic, accusing Russia of manipulating energy supplies for political gain. There was talk of sanctions against Moscow and of expelling Russia from the G-8 summit of world leaders.

Putin had to give in and restore the full flow of gas. But he remained fixated on punishing Ukraine and Poland. So he approached Schröder for permission and financing to build a double gas pipeline that would flow directly from Portovaya Bay near Vyborg in Russia to the German coast near Greifswald.[4] The Northern European Gas Pipeline (NEGP) would be 1,200 kilometers long, carrying 55 bcm of gas through its two pipes.[5]

When Russia first proposed the pipeline, democrats throughout Western Europe were skeptical and saw Putin's move for what it was: an attempt to cut out Poland and Ukraine from Russian gas deliveries to punish them for their independence and democratic leanings.

What other reason could Putin have had to build the pipeline? He claims he is just increasing the capacity for gas shipment, based on Gazprom estimates that "by 2010 the European gas market will need about

100 bcm of gas in excess of the existing long-term contracts." According to Gazprom, "the existing export gas pipelines from Russia to Europe will not be able to quench the growing gas thirst." So the NEGB is being built "to solve this problem."[6]

But an undersea pipeline is vastly more expensive than an overland pipeline. The obvious way to increase the capacity for gas export would be to build another pipeline right alongside the current right of way. But Putin chose to go the sea route instead. As the *Washington Post* speculated in its editorial: "The only possible reason for doing so was political: the Baltic Sea pipeline could allow Russia, a country that has made political use of its energy resources, to cut off gas to Central Europe and the Baltic states while still delivering gas to Germany."[7]

Gazprom hinted as much in its announcement of the NEGP project. "The North European gas pipeline . . . will begin to deliver Russian natural gas to Western Europe avoiding transit states along its route," i.e., Ukraine and Poland. "The [pipeline] will bypass transit states reducing sovereign risks and . . . enhancing reliability of export supplies."[8] In other words, the pipeline will permit Russia to cut off gas deliveries to Eastern European nations it doesn't like—with impunity.

In the 1950s and '60s, Russia used tanks to enforce its will in Eastern Europe. Now it is going to use energy deliveries.

So why would Gerhard Schröder go along with the plan? Again, the *Washington Post* nailed it in their editorial: "Many have wondered why Germany chose to go along with this project. Could it have been because the former chancellor realized that he was, in effect, creating his own future place of employment?"[9]

The news that private citizen Schröder had been unanimously selected as the chairman of the Shareholders' Committee of NEGP, shortly after Chancellor Schröder approved the pipeline, displeased his old friends. "It stinks," said Reinhard Buetikofer, cochairman of Germany's Greens, who were a coalition partner in Schröder's Social Democrat-led government.[10]

Gazprom blithely denied anything was amiss, saying that Schröder's new "position is not related to any kind of favor on our part."[11] One Gazprom official said, "Schröder was such an important figure that he was never going to have trouble finding a job."[12]

The *Washington Post* reported that "Schröder and the German-speaking

Putin built a strong political and personal alliance during the chancellor's seven-year tenure." The paper noted, "while other European governments have criticized Putin for squashing democratic institutions and freedoms in Russia, Schröder did not dwell on such concerns and focused instead on building German-Russian business and political ties." As the *Post* observed, Schröder had gone "out of his way" to ignore Putin's increasing suppression of freedoms in Russia and to "play down" Russia's brutal tactics in Chechnya. "Throughout his term in office," the paper said, "Mr. Schröder thwarted attempts to put unified Western pressure on Russia to change its behavior." [13]

Adding insult to injury, the chief executive of the pipeline consortium is Putin's close friend Matthias Warnig. The *Wall Street Journal* reported that Warnig was an officer in the Stasi, the East German secret police, and apparently met Putin when the future Russian president was based in East Germany, during his service in the KGB. [14]

The European Commission, the governing body of the EU, was so offended by the situation that it demanded that "Germany defend its decision to guarantee" a portion of the loan to build the pipeline. [15]

UPI reported that "Neelie Kroes, the commission's competition minister, wants an explanation as to why Berlin guaranteed part of the natural gas pipeline that will run beneath the Baltic Sea, from Russia to Germany, bypassing the Baltic States and Poland." [16]

The EU has the right to check out any "large state aid payments made by EU governments and can block them or demand repayment if they are deemed illegal." [17] A spokesman for Kroes said that "The commission requested information from the German authorities so that we can verify that any state support is fully compatible with the EU state aid rules." [18]

But Schröder's successor as German chancellor, Angela Merkel, blinked when it came to reversing the results of the extortion and Russian bribery. As leader of the Christian Democrats who opposed Schröder's Social Democrats, Merkel could have repudiated the agreement, and launched an investigation and even a prosecution of the former chancellor.

Instead, she chose to do nothing. She meekly accepted the pipeline deal, and pledged that Germany would meet its financial obligations under the agreement.

Why did Merkel show so little guts? The obvious answer is that she holds the chancellorship, and the Parliamentary coalition that sustains it, only by the sufferance of Schröder's party. Merkel's Christian Democrats—even when combined with their traditional allies, the Free Democrats—fell short of an absolute majority in the 2005 German general election. Though they outpolled the Socialist-Green Party coalition, they needed a third partner to govern. Unwilling to reach out to the Communists or the Green Party, they chose to form a "grand coalition" with their main rival, the Social Democrats. That decision left Merkel with no room to maneuver on the pipeline. Had she dissed the Putin-Schröder deal, she would have likely triggered the breakup of her coalition—and the fall of her government.

But there is a deeper reason behind the failure of the German political system to get tough with Schröder and make him account for his conduct. The fear of political instability hangs heavily over this land. Memories of their Nazi past loom large in German political circles. Once, Dick asked one of the leaders of the Christian Democrats why they did not demand a referendum on the proposed European Constitution, which Schröder and the Social Democrats had been pushing. "Referenda have not worked out too well for us in the past,"[19] he observed blandly, doubtless alluding to Hitler's seizure of power and his use of referenda to ratify it.

"Our task, as a political party," he explained, "is not just to seek power but to mediate public opinion so that it does not destabilize our system."[20] In other words, to give differing points of view a place to let off steam so that they don't rock the boat and endanger the democratic system. No American politician would give the least thought to the dangers of rocking the boat. No Republican pursuing Clinton, no Democrat chasing after Nixon or Bush, felt he was endangering the democratic system or the Constitution. Indeed, they probably felt they were *protecting* the Constitution and the rule of law. But in Germany you don't rock the boat too hard: They've been in the water before.

So Schröder got away with it, and Putin is busily proceeding with the pipeline construction. His device for blackmailing Eastern Europe will be in place by 2010—all because a German chancellor sold out the cause of freedom in return for a payoff. Now that's an outrage!

JACQUES CHIRAC'S LONG HISTORY OF PUBLIC CORRUPTION . . . AND HIS IMMUNITY FROM PROSECUTION

Jacques Chirac's political career is a testament to the adage: "Fool me once, shame on you. Fool me twice, shame on me." That the voters of France elected Chirac president in 1995 is a testament to his amazing ability to cover his tracks and avoid being snared by serious allegations of corruption. That they reelected him in 2001, after most of the charges had become public, is a tribute to the lax ethical standards in which France's political system wallows.

It is hard to imagine a modern American political figure who could succeed after being involved in so many scandals, all of which were well proven and blatant. It is even hard to conjure historical precedents that match Chirac's misdeeds. President Warren G. Harding or Louisiana governor Huey Long come to mind. Even President Richard Nixon's shenanigans pale by comparison.

Why isn't Jacques Chirac behind bars? Because a judicial decision in 1999 grants him almost total immunity from prosecution as long as he is president of France. It also shields him from having to testify about his role in corruption scandals, because the court found that having to do so would be "incompatible with his presidential functions."[21]

In December 2005, many of Chirac's henchmen were sentenced to serve time in prison for misdeeds that he most likely ordered or at least condoned. The defense attorney for one of the forty-seven people prosecuted in the massive corruption investigation spoke pointedly of the "empty chairs" at the trial, referring to the absent president of France, who hides behind the immunity conferred upon him by his office.[22]

His less fortunate accomplices include the folks listed in the sidebar.

CHIRAC'S PARTNERS IN CRIME[23]

- Michel Roussin, Chirac's chief of staff during his tenure as prime minister, was convicted of overseeing the collection of 2.5 million euros in bribes from firms anxious to be hired for school construction or repair work. Often acting as the bagman himself, he got a suspended jail term of four years and a fine of E35,000.

- Michel Giraud, a key ally of Chirac and former president of the Ile-de-France region, which embraces the suburban ring around Paris, was sentenced to four years of prison on probation and fined E80,000.

- Guy Drut, an Olympic track and field gold medal winner who served as Chirac's sports minister, was convicted of taking a phony job at a construction firm. Most of the illegal payments he collected went to Chirac's political party.

- Alaine Juppe, former prime minister and a key Chirac supporter, was found guilty in a case involving the misuse of city hall funds while Chirac was mayor of Paris. An appellate court later reversed Juppe's conviction, holding that he was prosecuted after the statute of limitations had expired.

- Didier Schuller, a former official with Chirac's Rally for the Republic party (RPR), was sentenced to five years in prison, three of them suspended, for "misuse of company assets." Maintaining that the blame should have rested on Chirac's shoulders, Schuller said his conviction was akin to blaming a flight attendant, not the pilot or the airline, for a plane crash.

- A total of forty-seven businessmen, public officials, and staff were prosecuted in the scandal. All but one were found guilty.

Chirac got off to an early start in his career of public sector crime. In 1969, just two years after he was first elected to the National Assembly, Chirac bought what the Guardian described as "a dilapidated sixteenth-century" residence, the Chateau de Bity." Purchased from money he had inherited from his father, the mansion came with twenty-seven acres of land. No sooner did he buy it than he got it listed as a historic monument, and "as such, a number of fat government subsidies rapidly became available to cover the renovation program he was planning." What's more, since it was an historic monument, the cost of the repairs provided so generous a tax deduction that in 1970 and 1971 he paid no income tax.[24]

The Chirac scandals first came to light in dramatic fashion in September 2000, when the French leftist daily newspaper Le Monde published a videotaped interview with businessman Jean-Claude Mery. He had made the videotape in 1997 "in case something happens to me." [25] *Le Monde* released the video one year after his death in 1999.

In the interview, Mery testified about his role in a scandal relating to the awarding of contracts for the construction of Parisian public housing projects. He said that he delivered five million francs (about US$1 million) in cash to Michel Roussin, Chirac's chief of staff, who was then serving as prime minister of France. Mery says he made the payment "in Chirac's presence" and noted that he took payoffs only "on Mr. Chirac's orders." [26]

The revelations shook French politics to its core. Investigative magistrate Eric Halphen led the investigation and summoned Chirac to testify in March 2001. But in September the Paris Appeal Court cancelled the summons, and took the "affair out of Judge Halphen's hands." [27]

After his removal, Halphen said, "I have proof the [bribery] scheme existed, and that the money went to Paris town hall and to the RPR (Chirac's Party). At the time the mayor of Paris was Jacques Chirac, the man who appointed the municipal officers who demanded the bribes was Jacques Chirac, and the head of the party that was the main beneficiary of the bribes was Jacques Chirac." [28]

Louise-Yvonne Cassetta, the RPR's former unofficial treasurer, testified that Chirac was "personally informed of all 'gifts' from building companies to the party." [29] François Ciolina, a former deputy director of the Paris public works department, called Chirac "the inspiration" for the kickback program. "Businessmen," he said, "knew that in order to win contracts, they had to deposit cash at the mayor of Paris' office." [30]

Yet Chirac, while refusing to testify, denies any knowledge of the kickback scheme and calls the charges "lies, calumny, and manipulations." [31]

Excuse our French, but how can Chirac have such sangfroid in the face of these serious charges? A predecessor as president of France, Valery Giscard d'Estaing, provided the answer: "Chirac can have his mouth full of jam, his lips can be dripping with the stuff, his fingers covered with it, the pot can be standing open in front of him. And when you ask him if he's a jam eater, he'll say: 'Me, eat jam, Never!'" [32]

But French political corruption is bipartisan. Apparently Mery turned the tape over to his friend, tax lawyer Alain Belot. Recognizing that the tape was political dynamite, Belot dangled it in front of Dominique Strauss-Kahn, then serving as finance minister in the Socialist government of Lionel

Jospin. His offer: He would turn over the tape if Strauss-Kahn agreed to reduce the back tax liability owed by Belot's famous client, the fashion designer Karl Lagerfeld.[33] Done! The tax bill was reduced and the tape made its way into Socialist hands for release at a pivotal time, right before Chirac's reelection campaign.[34] (Implicated in another financial scandal, Strauss-Kahn was subsequently forced to leave the Cabinet.[35])

Investigations revealed a network of corruption scandals that seemed to pervade almost every aspect of Chirac's regime as mayor of Paris. Forty businessmen and politicians were investigated for giving or taking kickbacks for high school construction and other projects in the Ile-de-France region. Contractors were asked to kick back 2 percent of the cost of the construction to the political parties, "either through legal contributions . . . or through cash deliveries [for the] fictitious employment of the staff of political parties or politicians."[36]

The largesse extended to both ends of the political spectrum. The RPR, of which Chirac was president, got 1.2 percent, while the rival French Socialist Party got 0.8 percent.[37] Prominent French companies including Sicra, Baudin Châteauneuf, GTM, Bouygues, Nord-France, Dumez, Chagnaud, and Fougerolles all admitted to making these payments.[38]

But Chirac, a presence who *Liberation Newspaper* said "towered" over the trial, was never convicted, or prosecuted, or even called to testify.[39]

One key stalwart of Chirac's party and his future interior minister, Charles Pasqua, was cited for a separate kickback scheme for work on public housing projects of the Hauts-de-Seine region.[40] Pasqua would later be fingered in documents recovered from Saddam Hussein's offices for accepting payoffs from the United Nations in the Oil-for-Food Program.

Former councilor Didier Schuller, accused of involvement in the public housing scandal, fled France and holed up in the Caribbean. When he returned to France in 2001, just as Chirac was announcing his candidacy for reelection, the president denied knowing him. " 'I have never heard of this affair,'" Chirac said, his leg bouncing nervously under the table. "I may have happened to be in the same place as him, but I didn't know him personally.' " The following day, *Le Monde* published a photograph of Chirac shaking hands with Schuller at City Hall.[41]

Schuller fled France after he was nabbed trying to get investigating judge Halphen disqualified from his case, according to the *International Herald*

Tribune, "by fabricating evidence that the judge's father-in-law had taken a bribe to influence the case's outcome." [42]

Schuller left France one step ahead of the gendarme to live in the Dominican Republic in apparent peace and security, knowing full well that the last thing Chirac wanted was to see him back in France. But Schuller's own son told the French media about his father's hiding place, and he came back to France to face trial.

Schuller's "core allegation depicts Mr. Chirac as the architect of a party machine in Paris City Hall that collected a percentage from contracts for subsidized municipal housing and public schools and amassed a multimillion-dollar political war chest." [43]

Under Chirac's tenure as mayor of Paris, city government employees (whose salaries were paid by taxpayer funds) were actually found to be working full-time for Chirac's RPR Party. American politics is often touched by scandals involving the use of public employees for political purposes, but this scandal was different. Those prosecuted were found to have been paid entirely by the city—while they spent every working moment employed by the party.

Chirac's signature is on many of the letters that led to these no-show jobs. As the *Guardian* points out, this suggests "that during his eighteen years as mayor he oversaw a huge fake jobs scam that benefited party workers, friends, and favor-seekers alike." The newspaper reports having seen a seventy-four-page printout of city hall staff in 1985 "that shows some three hundred people [with] salaries from L25,000 to L30,000 a year (about $100,000 at today's prices)" who were put on the payroll as consultants by Chirac's office. As the paper notes, however, "few had an office and only 45 are listed in the internal telephone directory for that year." Georges Quemar, a former head of personnel at city hall, says that the government employed "dozens and dozens of RPR activists who had a real job at the party HQ but were paid by the city hall." [44]

Those prosecuted included Juppe, who had served as secretary-general of the RPR from 1988–1995, while also being employed as deputy mayor of Paris in charge of the city's finances. [45]

The list of scandals goes on. Xaviere Tiberi, the spouse of Chirac's successor as mayor of Paris, was paid thirty thousand [BPS] for writing a thirty-six-page report that was replete with "spelling and grammatical mistakes" and was "extremely poorly written." The media speculated that it had been penned solely as a justification for a payment to her.[46]

In 2004, Parisian mayor Bertrand Delanoe filed a complaint saying that, under the RPR, government officials got 700,000 euros of free gardening services and public supplies from city employees.[47]

On June 28, 2001, Le Monde reported that Chirac, his wife, Bernadette, and his daughter Claude had used $300,000 in public funds to "buy airline tickets while he was Mayor of Paris." The junkets included a weekend trip for Chirac and his daughter to New York City on the Concorde, costing $16,000, all paid in cash.[48]

Chirac's defense to these charges is a gem. In effect, he admits paying cash for the tickets, but says that the money came not from bribes or kick-backs but from a $50 million government slush fund—a secret fund the president can spend as he wishes with no accountability.

Of course, since the president can have checks drawn on this account, Chirac's explanation raises the question of why he chose to pay for the airline tickets in cash.

But Chirac's rebuttal has another flaw: He wasn't president of France when the cash payments for tickets were made. He bought the tickets in question between 1992 and 1995, a period when he was neither prime minister nor president—and had no access to the slush fund. François Hollande, the chairman of the opposition Socialist Party at the time, said that he didn't "see how cash could have come from secret funds during that period, unless it was stocked up during a period of time."[49]

Even mundane contracts like the job of printing official notices for the Paris city government were sold to the highest bidder. In 1997, accusations surfaced that Chirac had favored the Sempap printing company through "fraud and favoritism" from 1986 to 1995, while he was mayor.[50]

In 2003, Eric Halphen, the investigating judge who was barred from probing into Chirac's role in these various financial scandals, published a book, Seven Years of Solitude, in which he describes how his privacy was invaded and his phone tapped during his investigation. He says he even came to fear for his own life. "I no longer believe in equal justice for all, because a lone man can triumph thanks to an organization nestled at the highest level of the state," Halphen wrote.[51]

So why isn't Chirac in jail? How did he get immunity and why wasn't he impeached? According to the *BBC*, "Under France's constitution, the president can be prosecuted during his term only for high treason and only by the High Court of Justice, which is called by Parliament and is made up of twelve deputies and twelve senators.[52]

"However, the constitution is silent on offenses committed before the president's accession to power, in Mr. Chirac's case when he was Paris mayor or heading the RPR."[53]

The Constitutional Council clarified the situation in 1999—by ruling that Chirac's immunity from prosecution extended even to acts committed before he became president. Then the compliant French courts went even further, ruling that not only could Chirac not be prosecuted, but he could not be summoned as a witness against his will.[54]

Why didn't the French Chamber of Deputies—its parliament—amend the constitution to strip Chirac of his immunity? Because Chirac controlled a majority in the chamber, effectively blocking any action. Even when the opposition Socialists took over the chamber, they were loathe to require the president's testimony or to lift his immunity. As noted, scandals in French politics are bipartisan!

The real question is why the voters of France reelected Chirac when his history of personal and political corruption had so recently been exposed. The answer seems to be that corruption just isn't that big a deal in French politics. Despite a Socialist campaign that harped on the president's scandals, Chirac finished first in the initial round of the French election. Jospin, his chief rival, didn't even make the runoff and finished third. In second place was the most reactionary politician in France, Jean-Marie Le Pen, who dismisses the Holocaust as a "detail of history."[55] Faced with a choice between Chirac and Le Pen, most voters held their noses and voted to reelect the scandal-ridden Chirac—by a margin of 82 to 18.

The key issue in the election, oddly enough, was crime. Chirac, who pledged a firm hand and an aggressive crackdown on delinquency, contrasted with Jospin's softer, more ideologically leftist approach. Ignoring the corruption that was staring them in the face, the French voted to reelect their president.

Why do the French work so hard to protect their president from investigation? One has only to contrast the French court decision holding that Chirac did not need to testify in the scandal that grew out of his administration of Paris with the United States Supreme Court decision that Paula Jones could proceed with her lawsuit against President Clinton, even while he stayed in office. In the worst example of judicial speculation, the Justices saw no reason that a private lawsuit would unduly burden the president or divert him from the duties of his office. (No, it just led to his impeachment and to a two-year paralysis of the U.S. government, that's all.)

There's some backstory here. French history may be far from the dark annals of Germany's past, but instability was rife under the Fourth Republic, which lasted from 1945 to 1958. Without a strong executive, and with premiers cobbling together flimsy majorities based on the shifting coalitions of numerous political parties, France became almost ungovernable. Premiers and their cabinets came and went, with an average tenure of only a few months per government.

Charles de Gaulle, who had led the Free French in World War II, had warned about allowing a government of parties to return to power after World War II (as we discuss in our book *Power Plays).* In 1957, all eyes turned to the former general as he led a virtual coup d'état, using the threat of military intervention by Algerian-based French paratroopers to dislodge the last Fourth Republic government. Taking power, de Gaulle drafted a new constitution with a strong president, whom the voters directly elected, rather than depending on parliamentary majorities.

But de Gaulle, who was often accused of wanting to be a dictator, had distinct Royalist tendencies. He spoke for a faction of French thinking that never really got over the ouster and execution of Louis XVI on the guillotine. In the nineteenth century, the Royalists repeatedly forced out republican governments in favor of monarchs. In the twentieth century monarchy fell out of fashion, but the weakness of parliamentary government—and its demoralizing collapse in the face of Nazi aggression in World War II—left a strong desire for a presidential office that could be described as imperial. It was in this tradition that de Gaulle and his successors governed. And Chirac sees no need to change.

ACTION AGENDA

Of course there's not much Americans can do about misgovernment in Germany and France. But we can and must understand that American standards of political integrity are not replicated on the European continent. The French and Germans, who could never understand why the Lewinsky scandal was such a big deal, simply don't share our perception of the importance of integrity in public office. While financial malfeasance is the be-all and end-all of American politics, it causes barely a ripple in France or Germany. Newspapers do their duty and report on the evidence, but the reality is that the public just doesn't seem to care.

In the recent controversy over the Bush decision to invade Iraq, Americans were shocked at the unfolding evidence of massive corruption in the United Nations' Oil-for-Food Program (see our chapter on the UN). How, we wondered, could we take seriously the votes of nations whose leaders had been so obviously and generously bribed to back the Iraqi dictator and derail any effort at regime change? But French public opinion didn't see it that way. To them, the corruption, while lamentable, did not seem to disqualify their leaders in the least—so fundamentally different is their attitude toward public corruption.

Yet there is a takeaway message here: from the debate over war with Iraq, to the corruption of the Oil-for-Food scandal, Americans have been mystified by the Europeans' willingness to allow selfish profit motives to tilt their moral compass, even in the direst political circumstances. Watching how they conduct their own politics, their behavior is, at least, a little easier to understand. After all, their own constituents don't seem to care when these politicians break the rules or use their offices for personal gain. Aren't we naive to expect them to behave any better on the world stage?

HOW SPECIAL INTEREST TRADE PROTECTIONISM ROBS US ALL

Every year, there's a lot of competition for Dumbest Statement of the Year. It was a particularly tough contest in 2006, when stupid public pronouncements spread like wildfire. But our vote is clear. CNN's Lou Dobbs deserves the award for this profundity: "Our consumers' addiction to cheap foreign imports has emboldened many of this country's leading political elites to further erode America's security and its sovereignty." [1]

Dobbs is supposed to know his way around business and financial issues. But he misses the essential point: imports are a consumer's best friend. Without them, the shelves of discount stores would be empty and we would be paying vastly more than we should for everything from clothing to computers.

As economist W. Michael Cox and business journalist Richard Alm have observed in a business analysis for the Federal Reserve Bank of Dallas, "Over the past five years, U.S. prices have actually fallen for a wide range of traded goods, such as computers, clothing, toys and photographic supplies. Most television sets now come from overseas, and their prices are down nearly 10 percent in the past five years. Americans pay 15 percent less for other video equipment and more than 25 percent less for computers and peripherals."[2] The sectors where costs have soared—such as health care and higher education—are ones where there is no foreign competition.

By holding down inflation, in other words, those "cheap foreign imports" which Dobbs derides make it possible for every American to have a higher standard of living at a lower cost.

But people who agree with Dobbs don't usually couch their arguments in terms of consumer prices—when it comes to stirring up voter passions, they know that kind of argument's a loser. So instead they talk about how we are exporting American jobs. Lobbyists plead with Congress, and with the Office of the United States Trade Representative, for protection against foreign imports, citing the families who'll be thrown out of work if their special interest measures don't pass.

Under the cover of protecting American jobs, however, big businesses and special interests—along with their Washington lobbyists—have found a way to foist punishing trade restrictions on American consumers. With minimal effects on the job market, these restrictions mean huge profits for lobbyists and their special-interest clients.

Cox and Alm have reviewed these subsidies, trade benefits, import quotas, and other favors our government grants to special interests to protect their workers against foreign competition, and have quantified exactly how much this protectionist policy has cost the American consumer versus how many jobs it has saved. The results show how phony the whole jobs protection argument really is.

Take sugar as an example. Because the United States imposes quotas on sugar imports from other countries—usually the poorest of our neighbors to the south—American consumers have to pay $1.8 billion more for everything from soda to candy bars to cakes to muffins to all other manner of sweet things. And how many jobs are saved? 2,261! That comes to a cost-per-job-saved of $826,104! The American sugar producers don't care about saving the jobs of the poor sugar workers in the United States. Trust us on that. They don't pay their workers anything approaching $826,104 per year. They push for the sugar quotas so that they can make an estimated $400 million of profits from the higher prices they can charge![3]

Why do Americans tolerate it? Because we don't know about it.

Or look at something as mundane as suntan lotion. As Cox and Alm report, trade protection against the importation of benzenoid chemicals, which are used in suntan lotion, cost about $300 million per year in higher prices for the end-product on the drugstore shelf. To justify these protectionism measures, which lead to this robbery from the American consumer, they say they save American jobs—all 216 of them! Saving those jobs cost us *$1.4 million for each worker.* Cox and Alm note that "it would cost far less to simply pay them not to work!"[4]

There are countless more examples. Import restrictions on frozen concentrated orange juice save 609 jobs, at a cost of almost $400 million in higher shelf prices. The cost per protected job? $635,000![5]

The following table, drawn from Cox's and Alm's analysis, shows the vast range of products that get special trade protection, and the tiny number of jobs they save. The third column indicates how much we are paying in higher prices for each American job that is being "saved." The point of this chart is that the assertions of the industries begging for protection—for their workers—is pure baloney. They're trying to use the workers as an excuse to maximize their profits.

THE COST OF PROTECTION[6]

Protected Industry	Jobs saved	Total cost (in millions)	Annual cost per job saved
Benzenoid chemicals	216	$297	$1,376,435
Luggage	226	$290	$1,285,078
Softwood lumber	605	$632	$1,044,271
Sugar	2,261	$1,868	$826,104
Polyethylene resins	298	$242	$ 812,928
Dairy products	2,378	$1,630	$685,323
Frozen concentrated orange juice	609	$387	$635,103
Ball bearings	146	$88	$603,368
Maritime services	4,411	$2,522	$571,668
Ceramic tiles	347	$191	$551,367
Machine tools	1,556	$746	$479,452
Ceramic articles	418	$140	$335,876

(continued)

Protected Industry	Jobs saved	Total cost (in millions)	Annual cost per job saved
Women's handbags	773	$204	$263,535
Canned tuna	390	$100	$257,640
Glassware	1,477	$366	$247,889
Apparel and textiles	168,786	$33,629	$199,241
Peanuts	397	$74	$187,223
Rubber footwear	1,701	$286	$168,312
Women's nonathletic footwear	3,702	$518	$139,800
Costume jewelry	1,067	$142	$132,870
Total	191,764	$44,352	
Average (weighted)			$ 231,289

Source: W. Michael Cox and Richard Alm, "The Fruits of Free Trade," Federal Reserve Bank of Dallas, 2002 Annual Report, page 19, available at http://www.dallasfed.org/fed/annual/2002/ar02b.html

Trade protection costs American consumers extra money that they can ill afford and doesn't succeed at protecting American jobs. According to Cox and Alm, for example, "subsidies to steel-producing industries since 1975 have exceeded $23 billion; yet industry employment has declined by nearly two-thirds."[7]

While tariffs—taxes on imports—have gone down, and now account for only about 2 percent of the average cost of an imported product, U.S. law uses a wide variety of nontariff measures to stem the flow of low cost imports for American consumers. These include quotas that limit the amount foreigners can sell to American importers; anti-dumping laws that tend to keep out low-cost products; licensing, labeling, and packaging restrictions; quality controls; domestic content laws that limit imports; and "voluntary" export restraints (where we twist a foreign country's arm and get it to agree to limit its sales to the United States).

We Americans have become accustomed to seeing these measures as good for our country—to root for exports and oppose imports. But it is *imports* that permit tens of millions of Americans to live middle class lives, thanks to their low costs.

Why do Congress and the U.S. Trade Representative's Office lend themselves to gouging the consumer in order to fatten the profits of these protected industries? Are they really falling for the phony argument about the job losses they are preventing? Not on your life. Congress uses trade protection legislation *to garner campaign contributions from each of these affected industries.*

If these industries sought no protection, congressmen and senators wouldn't be able to get their campaign contributions. The system rewards the profiteering producers, their lobbyists, and the congressmen who take their campaign donations—all at the expense of the average and unsuspecting consumer, who buys sugary cereal and frozen concentrated orange juice for tomorrow's breakfast.

For example, with its campaign contributions, the sugar industry buys huge influence in Congress. The U.S. law imposes quotas on foreign sugar imports, which raise prices in the United States and spread poverty in third world countries that depend on sugar for their economies—all in the name of saving 2,261 jobs.

In the 2004 election cycle, sugar political action committees donated $2,484,000 to candidates for Congress. The American Crystal Sugar Company Political Action Committee donated more than $850,000 to candidates, while the American Sugar Cane League of the USA gave almost half a million.[8]

These contributions are not wasted. As a result of the sweet largesse of the sugar industry, the Federal Reserve reports that "a small number of growers and refiners pocket an estimated $400 million a year."[9]

Thanks to the efforts of the sugar PACs and their compliant legislators, the cost of sugar on the American market is twenty cents per pound, which is more than double the world price of nine cents. American consumers pay the difference: $1.9 billion.[10]

CAFTA, the Central American Free Trade Agreement, really upset the sugar industry. The Bush administration negotiated the treaty to give the same free trade rights we give Mexico to the seven tiny nations of Central America and to the Dominican Republic. These generally desperately poor countries depend on sugar sales to survive, and American quotas were helping to keep them poor by restricting sugar imports so as to benefit a few hundred sugar beet farmers, mostly in Minnesota. Despite their efforts, however, Congress still passed CAFTA by two votes in the House of Representatives. But, Democrats, bought off by campaign contributions, over-whelmingly opposed the agreement.

And the sugar industry is far from unique. Every product and every crop has its henchmen in Congress who rely on campaign contributions—from their own particular special interests—to get them elected; then, in return, they promote legislation to benefit the industry. Pennsylvania's former senator Rick Santorum underscored the influence of agricultural special interests on the Agriculture Committees: "When I got on the Ag Committee," he has said, "I could just walk around the room and point to the crop that was represented by the senator who was there. I mean, they got on the committee to protect their crop." [11]

Every industry plays the same game: give campaign contributions and get protection in return, gouging the American consumer.

In the name of protecting 2,378 jobs, the dairy industry gave $2.3 million in campaign contributions to our politicians in the 2005–06 campaign cycle. [12]

Saying they were trying to save jobs in the lumber industry, timber PACs gave $831,000 in the 2005–06 cycle. [13] The total number of jobs saved was only 544.

Even the peanut industry got in on the act, giving $303,000 to protect its industry from foreign competition. [14] The number of jobs saved was 397.

Who pays the price? The American consumer. According to the Institute for International Economics, trade protectionism costs the average American household $6,027 each year. [15]

Perhaps worse, these trade restrictions punish the poorest countries in the world. With one-fourth of the world's wealth, but only 4 percent of its population, the U. S. domestic market has always been the key to helping third world countries emerge from poverty. Japan and Germany used this route to recover from the devastation of World War II; more recently, South Korea, Taiwan, Hong Kong, and Singapore have trod this well-worn path. Now, China, Thailand, Cambodia, India, and other countries are using their access to American consumers to get out of poverty.

But the process has not been one-sided. The fact is that foreign trade is a huge source of jobs for Americans. Thirty years ago, only about 11 percent of our economy came from foreign trade. Now it is up to almost one third.

The United States sells $1.3 trillion a year to the rest of the world—trade that has helped create almost 50 million jobs.[16] Once the Japans and Germanys and South Koreas of the world developed economically, they became excellent customers for the United States.

If the special interests' trade restrictions genuinely *produced* jobs in the United States, there would be a case for maintaining them. But all they do is provide a handful of jobs—while enriching a few special interests at the expense of the American consumer.

And, more than any other single factor, these trade restrictions promote global poverty. The recent round of global trade negotiations collapsed over the key issue of agricultural trade barriers.

Some of these trading barriers are quotas and other restrictions on the importation of foods and other products. But the most harmful, by far, are the subsidies western countries give to their farmers so they can undercut the prices of agricultural commodities on the world market. Farm subsidies in Europe amount to about 35 percent of the gross receipts of their farm sector, while they run to about 25 percent in the United States.[17] These subsidies strike directly at hundreds of millions of subsistence farmers in third world countries. Yet the number of jobs saved by these restrictions is minimal.

Indeed, the vast bulk of American farm subsidies go to the richest agribusinesses in the world. Agricultural subsidies are the most expensive form of domestic protection. As the *San Francisco Chronicle* noted, "From 1995 to 2002 the U.S. taxpayer doled out more than $114 billion to farmers, and in 2002 President Bush upped subsidies to $190 billion over the next 10 years."[18]

The Heritage Foundation reports that "these subsidy programs tax working Americans to award millions to millionaires and provide profitable corporate farms with money that has been used to buy out family farms." The Foundation reports that the cost of farm subsidies over a ten-year period comes to $4,400 for the average American household in tax payments and in increased food prices.[19]

By subsidizing agriculture, our government not only soaks us for more tax money but also makes it harder and harder for third world farmers to eke out a subsistence living. Truly the most selfish and greedy of our budgetary policies, farm subsidies cost the American taxpayer billions—and cut farm income in third world countries by $24 billion a year, according to the National Center for Policy Analysis.[20]

And who gets the benefit? Not America's family farms. The Environmental Working Group reports that 71 percent of the farm subsidies go to the richest 10 percent of farmers. And the so-called family farm hallowed in tradition? Sixty percent of farms get no subsidy at all,[21] while in 2002 seventy-eight farms got more than a million dollars in subsidies.[22] Further, the government subsidies are not spread out across the country; more than half of them go to only twenty-five congressional districts (out of 435).[23]

Most of the cash goes to subsidize five basic crops: corn (maize), soybeans, rice, cotton, and wheat. This chart shows how they divvy up the pie:[24]

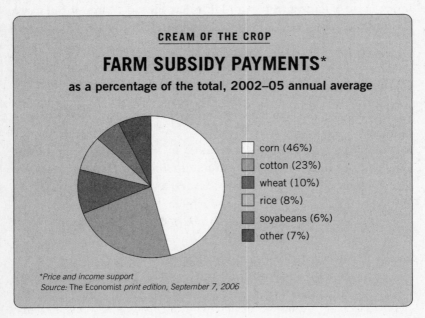

CREAM OF THE CROP

FARM SUBSIDY PAYMENTS*

as a percentage of the total, 2002–05 annual average

- corn (46%)
- cotton (23%)
- wheat (10%)
- rice (8%)
- soyabeans (6%)
- other (7%)

*Price and income support
Source: The Economist print edition, September 7, 2006

The big farmers sell their subsidy legislation to the country as a program to help the family farmers—but, as the Heritage Foundation points out, the more acres of land you farm, the more subsidy you get. So "large farms and agribusinesses—which not only have the most acres of land, but also, because of their economies of scale, happen to be the nation's most profitable farms—receive the largest subsidies. Meanwhile, family farmers with few acres receive little or nothing in subsidies. In other words, far from serving as a safety net for poor farmers, farm subsidies comprise America's largest corporate welfare program." [25]

Take a look at the sidebar for some of the so-called poor family farmers our tax dollars are supporting.

SUBSIDIES THEY DON'T DESERVE [26]

- David Rockefeller: the billionaire banker-philanthropist got $134,556 in 2001

- Scottie Pippen: the Portland Trailblazers basketball star got $25,315

- Ted Turner: the CNN founder walked off with $12,925

- Kenneth Lay: as he was fleecing Enron shareholders, he collected $6,019

- Bernie Ebbers: the former WorldCom CEO owned three farms that took in more than $4 million in Agriculture Department aid from 1998 through 2001 [27]

Source: Backgrounder #1542, Heritage Foundation, April 30, 2002, and USA Today, November 12, 2002

Isn't it nice to be reaching out to help such deserving folk as they work to toil the land?

Our farm subsidy programs, which tender a yearly payoff to a few farmers in a few places, plunges third world farmers into poverty.

Sometimes, "farmers" don't even have to farm their land to get a big subsidy—they can collect a check from the government for owning land and *not* farming it. As the *Washington Post* reports, "Since 2000, the U.S. government paid out $1.3 billion to 'farmers' who don't farm. . . . A Houston surgeon received nearly $500,000 for, literally, nothing." [28]

To pay for these agricultural subsidies, of course, the government sticks it to us, the taxpayers. The subsidies drive up prices, sucking money from our consumers. And our trade policies impoverish much of the third world. All in the name of saving a handful of American jobs.

In today's global economy, the line between foreign and domestic manufacturing is blurred. A 2001 Ford Escort, allegedly an American car, was only 60 percent produced in the United States, while a 2001 Honda Civic, seemingly a foreign car, was 75 percent made in the USA.[29]

By 2001, only about half of the cars sold in the United States were made in America by American firms. But an additional one-fourth were made by foreign firms who manufactured in the States, with American workers. Only about a quarter of our cars were foreign made by foreign companies.[30]

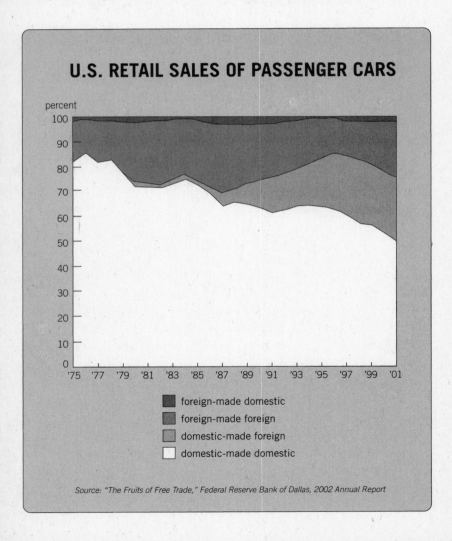

U.S. RETAIL SALES OF PASSENGER CARS

- foreign-made domestic
- foreign-made foreign
- domestic-made foreign
- domestic-made domestic

Source: "The Fruits of Free Trade," Federal Reserve Bank of Dallas, 2002 Annual Report

ACTION AGENDA

Special interests seeking to pad their corporate profits are abusing American consumers and taxpayers, and it won't end until the American consumer speaks out! We've all been conned by the claim that we're protecting our fellow American workers, when all we've *really* been doing is protecting a handful of jobs and billions in corporate profits—with both higher taxes and higher prices.

When Congress considers trade legislation or the farm subsidy bills, most Americans turn the channel, uninterested in the technical details of such legislation. But if we realized that every time we drink orange juice, eat a cookie, buy shoes, build a home, apply cosmetics, buy a quart of milk, eat a bag of peanuts, purchase a pocketbook, or pay for practically anything, we're paying for these trade subsidies, maybe we'd wake up and start paying attention.

CONCLUSION

It may seem counterintuitive, but outrage is actually an optimistic reaction. It's the opposite of cynicism. The same events and situations that kindle outrage in the optimistic create only alienation, apathy, and disgust in the pessimistic. If we don't believe things could be better, we're not outraged by imperfection. Bad things just confirm our predictions, and once we've lost hope that they can ever improve, they push us further into despair and inaction.

Even in our outrage, though, we Americans are an optimistic bunch, fully committed to the belief that things can get better. So we hope the anger we've summoned in this book will catalyze civic involvement, voting, and political participation. To channel that activism constructively, we hope our suggestions for action will stir some ideas—and action—of your own. There's no need to sit back and watch things get worse when we can improve them by our own energy and commitment.

There is some good news. Even as we've been working on this book, some of the situations we've written about have improved. State Farm is no longer stiffing its Mississippi policyholders devastated by Katrina. The House and Senate are moving to cure some of the abuses—though far from all—that we decry in our chapters on Congress. Student loan interest rates seem to be coming down, and the minimum wage may be going up. You see? Progress is possible.

But there's been some backsliding, too. President Bush has been forced to abandon his national security surveillance program, and the drug companies and teachers unions are still getting away with the fleecing of sick people and children. Illegal immigrants, having overstayed their visas, are even less likely to be caught than they were a year ago.

So we've got to act—every one of us. We must fight and fight and fight to see these outrages reversed and our country put back on the right track. Complacency will only lead to corruption. Activism will make a difference.

ACKNOWLEDGMENTS

Our heartfelt thanks—and our hearts—go out to Judith Regan, our former publisher, with whom this was our sixth book. Her brilliance, creativity, and intuitive grasp have been something to witness and we will both miss her very much.

Fortunately, the other half of the tandem with which we have worked so well, Calvert Morgan, is still there and did so very much to make this book what it is. He is a joy to work with and a good friend and we both thank him very much.

Thanks to Katie Vecchio for her research, editing, and proofing. She guided us through the minefield of research and data with great skill.

And to Thomas Gallagher for making sense of the hundreds and hundreds of footnotes, often maddeningly scattered in the early drafts in the wrong places.

Finally, we want to thank Cassie Jones, Donna Lee Lurker, Brittany Hamblin, Gregg Sullivan, and Suzanne Wickham at HarperCollins for their wonderful help in producing this book.

NOTES

INTRODUCTION

1. Interior Minister, Charles Pasqua: Robin Wright and Colum Lynch, "Hussein Used Oil to Dilute Sanctions," *Washington Post,* October 7, 2004, http://www.washingtonpost.com/wp-dyn/articles/A13313-2004Oct6.html.
2. between $1–3.5 million: Ibid.
3. who took $156,000: Francis Harris, "France's Ambassador for Life Admits Role in the Oil for Food Scandal," *The Telegraph,* November 18, 2005.
4. Chief of Staff to Vladimir Putin: "U.S. Senate Links Kremlin to Iraqi Oil-for-Food Scandal," *Mosnews.com,* May 16, 2005, http://www.mosnews.com/news/2005/05/16/oilforfoodrussia.shtml.
5. 13 million barrels: Pasqua: Wright and Lynch, "Hussein Used Oil."
6. $2500 each month: Claudia Rosett, "Annan's Son Took Payments Through 2004," *New York Sun,* November 26, 2004.
7. $114,000 monthly pension: Terence O'Hara, "Study of Fannie Mae Cites 'Perverse' Executive-Pay Policy," *Washington Post,* March 31, 2005.
8. more than $25 million: Opinion Editorial, "Fannie Mae's Fall From Grace," *Rocky Mountain News,* August 13, 2006, http://www.rockymountainnews.com/drmn/editorials/article/0,2777,DRMN_23964_4911017,00.html.
9. a $1,932,000 bonus: Albert B. Crenshaw, "High Pay at Fannie Mae for the Well-Connected," *Washington Post,* December 23, 2004.
10. "no conflict of interest": Jodi Rudoren, "Lawmaker and Beneficiary Bought a Farm," *New York Times,* April 26, 2006, http://www.nytimes.com/2006/04/26/washington/26farm.html?ex=1165554000&en=648392be1ad78cee&ei=5070.
11. $6.3 and 24.9 million: Chris Cillizza, "W. Va.: 12-Term Democrat at Center of Ethics Storm," *Washington Post,* April 13, 2006.
12. "he had never seen": Mark Leibovich, "Washington Traffic Jam? Senators-Only Elevator," *New York Times,* August 2, 2006, http://www.nytimes.com/2006/08/02/washington/02elevator.html?ex=1312171200&en=0e6c2360190e321d&ei=5088&partner=rssnyt&emc=rss.

13. "I hesitate to say": Ibid.

14. hostile glares.": Ibid.

15. "all or nothing": Carl Hulse, "Wage Bill Dies; Senate Backs Pension Shift," *New York Times,* August 4, 2006, http://www.nytimes.com/2006/08/04/washington/04cong.html?ex=131234 4000&en=a1a2f997bb6072b7&ei=5088&partner=rssnyt&emc=rss.

16. 6.6 million workers: "Minimum Wage Facts at a Glance," Economic Policy Institute, November 2006, http://www.epinet.org/content.cfm/issueguides_minwage_minwagefacts.

17. high school diploma or less: Patrik Jonsson, "If Minimum Wage Is Raised, Who Benefits?" *Christian Science Monitor,* June 22, 2006, http://www.csmonitor.com/2006/0622/p01s03-usec.html.

18. the next fifteen years: Joel Friedman and Ruth Carlitz, "Estate Tax Reform Could Raise Much-Needed Revenue: Some Reform Options with Low Tax Rates Raise Very Little Revenue," Center on Budget and Priorities, March 16, 2005, http://www.cbpp.org/3-16-05tax.htm.

19. liable for the tax: Jeffrey H. Birnbaum, "Senate Plan to Repeal Inheritance Tax Fails," *Washington Post,* June 9, 2006, http://www.washingtonpost.com/wp-dyn/content/article/2006/06/08/AR2006060800138.html.

20. subject to the estate tax: Conor Kenny and Taylor Lincoln, "Spending Millions to Save Billions, The Campaign of the Super Wealthy to Kill the Estate Tax," *Public Citizen,* April 2006, p. 8, http://www.citizen.org/documents/EstateTaxFinal.pdf#search=%22Spending%20Mil lions%20to%20Save%20Billions%2C%20The%20Campaign%20of%20the%20Super%20 Wealthy%20To%20Repeal%20the%20Estate%20Tax%22.

21. $3.5 million in 2009: David Cay Johnston, "A Boon for the Richest in an Estate Tax Repeal," *New York Times,* July 8, 2003, http://www.nytimes.com/2006/06/07/business/07impact .html?ex=1307332800&en=222cd679ecc077fc&ei=5088&partner=rssnyt&emc=rss.

22. spent over $200 million: Charlotte Sector, "End of Campaign for 'Paris Hilton Tax Break,'" ABC News, June 8, 2006, http://abcnews.go.com/Politics/story?id=2049451&page=1.

23. "The long-running": Joan Claybrook, "Wealthy Families' Campaign to Repeal Estate Tax Is Big Con Job," *Public Citizen,* April 25, 2006, http://www.citizen.org/pressroom/release.cfm? ID=2181.

24. estate tax of $4.8 billion: Kenny and Lincoln, "Spending Millions to Save," p. 11.

25. more than $4 million: Ibid., 27.

26. "a near-record low": Lydia Saad, "Congress Approval at Twelve Year Low," *Gallup* News Service, April 17, 2006, http://www.galluppoll.com/content/?ci=22435.

CHAPTER ONE

1. visa overstays account: Jeffrey S. Passel, "Size and Characteristics of the Unauthorized Migrant Population in the U.S.," Pew Hispanic Center, March 7, 2006, http://pewhispanic.org/reports/report.php?ReportID=61.

2. Tourists 74 percent: Elizabeth M. Grieco, "Temporary Admissions of Nonimmigrants to the United States: 2005," *Department of Homeland Security Annual Flow Report,* July 2006, http://www.dhs.gov/xlibrary/assets/statistics/publications/2005_NI_rpt.pdf.

3. legally admit 33 million: Jessica Vaughan, "Shortcuts to Immigration: The 'Temporary' Visa

Program Is Broken," Center for Immigration Studies, January 2003, http://www.cis.org/articles/2003/back103.html.

4. "in 2001, more than": Vaughan, "Shortcuts to Immigration."

5. "they've only got one": Spencer S. Hsu, "Legal Residents Face Fingerprinting at Ports: Program Requires More Proof of Identity," *Washington Post,* July 28, 2006, http://www.washingtonpost.com/wpdyn/content/article/2006/07/27/AR2006072701452.html.

6. "We can't handle": Author conversations with President Bill Clinton, 1995 and 1996.

7. "four former consular officers: Joel Mowbray, "Visas for Terrorists: They Were Ill-Prepared. They Were Laughable. They Were Approved," *National Review,* October 28, 2002, http://www.findarticles.com/p/articles/mi_m1282/is_20_54/ai_92712403.

8. "All six experts strongly": Ibid.

9. Most of the information on the hijackers' immigration status comes from a Federation for American Immigration Reform (FAIR) report entitled "Identity and Immigration Status of 9/11 Terrorists," available on the web at http://www.fairus.org/site/PageServer?pagename=iic_immigrationissuecentersc582.

10. "applied together for travel: Joel Mowbray, "Visas That Should Have Been Denied: A Look At 9/11 Terrorists' Visa Applications," *National Review,* October 9, 2002, http://www.nationalreview.com/mowbray/mowbray100902.asp.

11. "claimed to be a student": Ibid.

12. "ALQUDOS HTL JED": Ibid.

13. There were 1.5 million: Steven A. Camarota, "The Muslim Wave: Dealing with Immigration from the Middle East," *National Review,* September 16, 2002, http://www.findarticles.com/p/articles/mi_m1282/is_17_54/ai_90888287.

14. "Even after the terror attacks": Ibid.

15. in 2001, 573,000 people: Vaughan, "Shortcuts to Immigration."

16. "By all accounts fraud": Ibid.

17. "Even if the intrepid": Ibid.

18. "In an incredible": Ibid.

19. "more than 10,300 schools": Jason Ryan, "FBI Seeks Missing Egyptian Students," *ABC News,* August 9, 2006, http://abclocal.go.com/ktrk/story?section=nation_world&id=4444674.

20. "Michael O'Hanlon said": Ibid.

21. "Without exit checks": "Misplaced Border Priorities: US-Visit Program Must Track Exits, Too," *Newsday,* August 1, 2006.

22. "millions of legal": "Possible expansion of fingerprinting at U.S. port of entries," *usagreen cardcenter.com,* August 22, 2006, http://www.usagreencardcenter.com/news.htm.

23. Of the thirty-three million: Vaughan, "Shortcuts to Immigration."

24. Table 1. Nonimmigrant Visas (NIVs): Ibid.

25. More than a quarter (27%): Editorial, "The Butchers' End," *Washington Times,* August 12, 2006, http://www.washtimes.com/op-ed/20060811-090001-3831r.htm.

26. "bloody 18th Street Gang ": Heather MacDonald, "Crime & the Illegal Alien: The Fallout from Crippled Immigration Enforcement," Center for Immigration Studies, June, 2004, http://www.cis.org/articles/2004/back704.html.

27. has only 2,000 officers: Ibid.

28. "in cities. . . . the police": Ibid.

29. "officers can't legally question": Juan A. Lozano, "Critics: Policy Makes Houston Haven for Immigrants," Associated Press, July 30, 2006, http://oilpatchdemocrats.blogspot.com/2006/07/help-this-evening-if-you-can.html.

30. "police need immigrants": Ibid.

31. "Orange County . . . they actively": Heather MacDonald, "ICE, ICE Baby: Orange County Provides the Bush Administration with a Significant Enforcement Opportunity," *National Review*, August 7, 2006, http://article.nationalreview.com/?q=YjRiYzRlYTM4ZjlhOThjY mRlNDUzNjQ3NjRmY2RlZGY.

32. "USCIS faces the continuing": Richard L. Skinner, "USCIS Faces Challenges in Modernizing Information Technology," *Department of Homeland Security*, September, 2005, http://www.dhs.gov/xoig/assets/mgmtrpts/OIG_05-41_Sep05.pdf.

33. "draw a yawn": MacDonald, "Crime & the Illegal Alien."

34. "an elaborate set": Ibid.

35. "A regular immigration": Ibid.

36. "lack of resources": Ibid.

37. "the process immediately bogged": Ibid.

38. "Illegal aliens have": Chelsea Schilling, "Illegal Aliens Linked to Gang-Rape Wave," *WorldNetDaily.com*, August 22, 2006, http://www.worldnetdaily.com/news/article.asp?ARTICLE_ID=51424.

39. "A failure that puts": Ibid.

40. "Deportations by Year": "Table 40: Aliens Expelled, Fiscal Years 1892–2004," Yearbook of Immigration Statistics: 2004, Department of Homeland Security.

41. It is estimated: "U.S. Resume Deportations to Haiti," MyCaribbeanNews.com, August 14, 2005, http://www.mycaribbeannews.com/inter/050414d.html.

42. "to find, arrest, and place": News release, "ICE Adds Seven New Fugitive Operations Teams To Its Nationwide Arsenal," *U.S. Immigration and Customs Enforcement*, August 9, 2006, http://www.hooyou.com/news/news081006ice.html.

43. "had convictions for crimes": Ibid.

44. "As a result, tens": MacDonald, "Crime & the Illegal Alien."

45. "even after September 11": Camarota, "The Muslim Wave."

46. the state sponsors of terrorism: Office of the Coordinator for Counterterrorism, "State Sponsors of Terrorism," U.S. Department of State, 2006, http://www.state.gov/s/ct/c14151.htm.

47. safe havens for terrorists: "State Department's Annual Report on Terrorism," *U.S. Department of State*, http://www.state.gov/documents/organization/65466.pdf.

CHAPTER TWO

1. "was given the right": Claudia Rosett, "The Oil-for-Food Scam: What Did Kofi Annan Know, and When Did He Know It?" *Commentary*, http://www.defenddemocracy.org/publications/publications_show.htm?doc_id=228669.

2. UN collected $1.9 billion: Ibid.

3. $4.4 billion to France: Ryan Balis, "Support for United Nations Justifiably Weakened by Financial, Sex and Human Rights Scandals," The National Center for Public Policy Research, July, 2006, http://www.nationalcenter.org/NPA545UNScandals.html.

4. "the Secretariat was": Rosett, "The Oil-for-Food Scam."

5. "There was no disclosure": Ibid.

6. "may have begun covertly": Ibid.

7. The Swiss firm, Cotecna: Paul A. Volcker, "Independent Inquiry Committee into the United Nations Oil-for-Food Programme," www.iic-offp.org, October 27, 20005, http://www.foxnews.com/projects/pdf/final_off_report.pdf.

8. "as oil-for-food": Claudia Rosett, "Oil-for-Terror?" *National Review,* April 18, 2004, http://www.nationalreview.com/comment/rosett200404182336.asp.

9. "Conlon said he": Eric Shawn, "Early Warning Not Heeded on Oil-for-Food," Fox News, September 21, 2004, http://www.foxnews.com/story/0,2933,132970,00.html.

10. "for example, we gave": Ibid.

11. "It worked like this": Rosett, "The Oil-for-Food Scam."

12. "Benon Sevan, the former": Balis, "Support for United Nations."

13. "solicited money from": Ibid.

14. Rev. Jean-Marie Benjamin: Volcker, "Independent Inquiry Committee."

15. "Corruption of the program": Balis, "Support for United Nations."

16. "firsthand knowledge it acquired": Ibid.

17. "There were provisions": Balis, "Support for United Nations."

18. "transportation fees": Ibid.

19. "Some items of questionable": Rosett, "The Oil-for-Food Scam."

20. "both Sevan and Annan": Rosett, "The Oil-for-Food Scam."

21. "against the resistance": Ibid.

22. "Olympic sport city": Ibid.

23. The list included former: Ibid.

24. "two firms": Ibid.

25. "One link ran from": Rosett, "Oil-for-Terror?"

26. "This Liechtenstein trust": Ibid.

27. "was a subsidiary": Ibid.

28. "had arranged to receive": Associated Press, "Korean Businessman Guilty in Oil-for-Food Case," Fox News, July 13, 2006, http://www.foxnews.com/story/0,2933,203377,00.html.

29. "The Aug. 28, 1998 fax": "Documents Challenge Kojo Annan's Story," Fox News, December 16, 2004, http://www.foxnews.com/story/0,2933,141656,00.html.

30. "K Annan, United Nations": Ibid.

31. Colum Lynch, "Kofi Annan Cleared in Corruption Probe," Washington Post, March 30, 2005, http://www.washingtonpost.com/wp-dyn/articles/A9495-2005Mar29.html.

32. "may have earned": Ibid.

33. "participating in two meetings": Ibid.

34. "he acknowledged the encounters": Ibid.

35. "Kojo Annan actively participated": Lynch, "Kofi Annan Cleared."

36. "In hindsight, I can say": Ibid.

37. seven months to complete: Claudia Rosett and George Russell, "UN Family Ties: Is There a Replay of the Kofi and Kojo Annan Scandal?" Fox News, June 20, 2005, http://www.foxnews.com/story/0,2933,160081,00.html.

38. "comes as a great relief": Lynch, "Kofi Annan Cleared."

39. The United Nations deploys: United Nations Dept. of Public Information, "UN Peacekeep-

ing Missions," *Information Please® Database,* http://www.infoplease.com/ipa/A0862135 .html.

40. 221 investigations of sexual improprieties: Balis, "Support for United Nations."

41. "the UN's Abu Ghraib": Jonathan Clayton and James Bone, "Sex scandal in Congo threatens to engulf UN's peacekeepers: They should be rebuilding the country, but foreign workers face serious accusations . . . scandal . . . threatens to become the UN's Abu Ghraib," *The Times,* December 23, 2004, http://www.timesonline.co.uk/tol/news/world/article405213 .ece?token=null&offset=12.

42. "seeking to obstruct": Colum Lynch, "UN Sexual Abuse Alleged in Congo Peacekeepers Accused in Draft Report," *Washington Post,* December 16, 2004, http://www.washingtonpost .com/wp-dyn/articles/A3145-2004Dec15.html.

43. "alleged rape, prostitution": Ibid.

44. Sexual exploitation and abuse: Ibid.

45. "inappropriate conduct": Ibid.

46. "The fact that": "The New World Disorder: U.N. 'peacekeepers' rape women, children," *World Net Daily,* December 24, 2004, worldnetdaily.com.

47. "I am afraid": Ibid.

48. "The situation appears": Ibid.

49. "If a soldier is found guilty": Balis, "Support for United Nations."

50. "The former secretary-general": Staff reporter, "UN Chief Helped Rwanda Killers Arm Themselves," *The Observer,* September 3, 2000, http://observer.guardian.co.uk/inter national/story/0,6903,363732,00.html.

51. "helped launder the international aid": Linda Melvern, *A People Betrayed: The Role of the West in Rwanda's Genocide* (Zed Books, 2000), http://findarticles.com/p/articles/mi_ m1316/is_12_32/ai_68148605.

52. "the deal was": "UN chief helped Rwanda killers arm themselves," *The Observer* (London), September 3, 2000.

53. "Asked about the wisdom": Staff reporter, "UN Chief Helped Rwanda."

54. killed at least 300,000 Rwandans: "Interview with Boutros Boutros-Ghali," *PBS: Frontline,* January 21, 2004, http://www.pbs.org/wgbh/pages/frontline/shows/ghosts/interviews/ghali .html.

55. "one of my greatest failures": Ibid.

56. "federal investigators have now": Claudia Rosett and George Russell, "The UN's Spreading Bribery Scandal: Russian Ties and Global Reach," Fox News, September 07, 2005, http:// www.foxnews.com/story/0,2933,168591,00.html.

57. "held various portfolios": Ibid.

58. "in significant financial misconduct": Ibid.

59. "not sufficiently supported": Eric Shawn, "Report Details Deficiencies in U.N. Procurement Department," Fox News, December 6, 2005, http://www.foxnews.com/story/0,2933, 177817,00.html.

60. "UN procurement employees": Ibid.

61. "There does not appear": Ibid.

62. "an open invitation": Geoff Earle, "UN Fraud Is Building," *New York Post,* June 21, 2006, http://www.un.org/cmp/uncmp/news/2006-06-21_NYP.pdf#search=%22UN%20FRAUD %20IS%20BUILDING%22.

63. "is an evergreen shrub": Joseph Goldstein, "U.N. Employee Is Charged with Drug Smuggling," *New York Sun,* July 27, 2006, http://www.nysun.com/article/36814.

64. "an important cog":Ibid.

65. "to be frank": "Testimony by Ambassador Bolton before the Senate Foreign Relations Committee," May 25, 2006, www.reformtheun.org/index.php?module=uploads&func=download&fileId=1512.

66. "changing that culture": Ibid.

67. "would have [made it] irresponsible": Ibid.

68. "beholden to those": Ibid.

69. "one of the keys to UNICEF's": Ibid.

70. "creates an entirely different": Ibid.

71. "these countries have": Ibid.

72. "the Declaration's 30 articles": Balis, "Support for United Nations."

73. "It is a regrettable yet": Ibid.

74. "When this great institution's": Ibid.

75. "The Commission reached": Joseph Loconte, "The United Nations and Human Rights: A Strategy for Meaningful Reform," The Heritage Foundation, April 19, 2005, http://www.heritage.org/Research/Religion/tst041905a.cfm.

76. "part of an effort": United Press International, "U.N.'s Hypocrisy on Rights Commission Jolts U.S.," Newsmax.com, May 5, 2001, http://www.newsmax.com/archives/articles/2001/5/4/214951.shtml.

77. "The U.S.,": Staff, "U.S. Funding Debated After Ouster from U.N. Human Rights, Drug Panels," *Baptist Press,* May 9, 2001, http://www.bpnews.net/bpnews.asp?ID=10855.

78. "commitment to human rights": Barbara Crossette, "For First Time, U.S. Is Excluded from UN Human Rights Panel," *New York Times,* May 4, 2001, http://select.nytimes.com/search/restricted/article?res=F40F10F6385C0C778CDDAC0894D9404482.

79. "banned chemical weapons": Jean-Claude Buhrer, "Human Rights on the rocks at the UN," *Reporters Sans Frontières,* April 22, 2004, http://www.rsf.org/imprimer.php3?id_article=9912.

80. "Chinese ambassador choked": Ibid.

81. "With this vote, every": Barbara Crossette, "China Maneuvers to Avoid Debate on Its Rights Record in UN," *New York Times,* April 19, 2001, http://select.nytimes.com/gst/abstract.html?res=F20610FB3F540C7A8DDDAD0894D9404482&n=Top%2fNews%2fWorld%2fCountries%20and%20Territories%2fChina.

82. "some of the world's most": Mark P. Lagon, "Advancing the Promotion and Protection of Human Rights at the United Nations," U.S. Department of State, October 29, 2005, http://www.state.gov/p/io/rls/rm/55840.htm.

83. "days after Fidel Castro": Mark Falcoff, The United Nations Human Rights Commission, *The American Enterprise Institute for Public Policy Research,* May 7, 2003, http://www.aei.org/publications/filter.all,pubID.17167/pub_detail.asp.

84. "five members of the commission": Barbara Crossette, "A UN Paradox: Some on Rights Panel Are Accused of Wrong," *New York Times,* March 13, 2001, http://select.nytimes.com/gst/abstract.html?res=F20714F93B5E0C708DDDAA0894D9404482&n=Top%2fReference%2fTimes%20Topics%2fOrganizations%2fU%2fUnited%20Nations%20.

85. 30 percent of the: Encyclopedia excerpt, "Israel and the United Nations," *Wikipedia.org,* http://en.wikipedia.org/wiki/Israel_and_the_United_Nations.

86. "tyrannical regimes such as": Nile Gardiner and Brett D. Schaefer, "John Bolton: A Powerful Voice for America at the United Nations," The Heritage Foundation, July 26, 2006, http://www.heritage.org/Research/InternationalOrganizations/wm1179.cfm.

87. "by adopting a politicized resolution": Press release, "Lebanon/Israel: UN Rights Body Squanders Chance to Help Civilians," Human Rights Watch, August 11, 2006, http://hrw.org/english/docs/2006/08/11/lebano13969.htm.

88. "could . . . replicate": Testimony by Ambassador Bolton before the Senate Foreign Relations Committee, "Challenges and Opportunities in Pushing Ahead on UN Reform," May 25, 2006, United States Mission to the United Nations, press release, May 25, 2006, http://www.usunnewyork.usmission.gov/06_126.htm.

89. "Is life so dear . . . ," *Quote DB,* http://www.quotedb.com/quotes/2357.

90. Annual Contributions to the United Nations Budget: *United Nations: 2006 Status of Contributions to the Regular Budget, International Tribunals, Peacekeeping Operations and Capital Master Plan.*

CHAPTER THREE

1. "A director may publicly": Stephanie Strom, "A.C.L.U. May Block Criticism by Its Board," *New York Times,* May 24, 2006, http://www.commondreams.org/headlines06/0524-06.htm.

2. "For the national board": Ibid.

3. "You sure that didn't": Ibid.

4. "The mission of the ACLU": Mission Statement, American Civil Liberties Union, http://www.aclu.org/about/index.html.

5. "We would never terminate": Adam Liptak, "A.C.L.U. Board Is Split Over Terror Watch Lists," *New York Times,* July 31, 2004, http://www.refuseandresist.org/police_state/art.php?aid=1467.

6. his "clever" interpretation: Ibid.

7. "Do we do more harm": Ibid.

8. "If I give the A.C.L.U.": Stephanie Strom, "A.C.L.U.'s Search for Data on Donors Stirs Privacy Fears," *New York Times,* December 18, 2004, http://www.nytimes.com/2004/12/18/national/18aclu.html?ei=5088&en=bf813c01c8c9ea38&ex=1261026000&partner=rssnyt&pagewanted=print&position=.

9. "It goes against A.C.L.U.": Ibid.

10. "For centuries, bankers": ACLU statement, "Privacy & Technology: Consumer Privacy," American Civil Liberties Union, http://www.aclu.org/privacy/consumer/index.html

11. "where he chastised her": Strom, "A.C.L.U. May Block Criticism."

12. "Anthony went on": Ibid.

13. "told me that he would": Ibid.

14. Attorney General's office called the ACLU: Stephanie Strom, "ACLU Warned on Rules to Limit Members Speech," *New York Times,* June 19, 2006, http://www.nytimes.com/2006/06/19/us/19aclu.html?ex=1308369600&en=507e8c82c537e9fa&ei=5088&partner=rssnyt&emc=rss.

15. "Like the proposal governing": Wendy Kaminer, "How the ACLU Lost Its Bearings," *Los Angeles Times,* July 2, 2006, http://www.latimes.com/news/opinion/sunday/commentary/la-op-kaminer2jul02,0,4254534.story?coll=la-sunday-commentary.

16. "kiss my Hispanic ass": Josh Gerstein, "For ACLU's Anthony Romero, These Should Be the Best of Times," *New York Sun,* June 27, 2006, http://www.nysun.com/article/35101.

17. "kiss my Puerto Rican ass": Ibid.

CHAPTER FOUR

1. in session for 108 days: Kathy Kiely, "Lawmakers Get Out of the House," *USA Today,* March 20, 2006, http://www.usatoday.com/news/washington/2006-03-19-house-session_x.htm.

2. To a weekend time: Lindsey Layton," Culture Shock on Capital Hill: House to Work Five Days a Week, *Washington Post,* December 6, 2006, http://www.washingtonpost.com/wp-dyn/content/article/2006/12/05/AR2006120501342.html.

3. "There is a very": "House Democrats ready to Start '100 hour' clock," CNN.com, January 8, 2007, http://www.cnn.com/2007/POLITICS/01/08/congress.democrats.reut/index.html.

4. House Non-Session Day Schedule for 2007: "Days in Session Calendars, 110th Congress 1st Session," The Library of Congress, http://thomas.loc.gov/home/ds/h1101.html.

5. Senate schedule for 2007: Ibid.

6. "No man's life": Gary Galles, "In Praise of Irreverence," The Ludwig von Mises Institute, June 14, 2002, http://www.mises.org/story/983.

7. Accomplishments of the 109th Congress: Every piece of legislation listed can be found in the roll call votes, available at http://thomas.loc.gov/home/rollcallvotes.html.

8. almost five million dollars: Report, "Join Congress-See the World: Part II—Privately Funded Travel," http://politicalmoneyline.com/cgi-win/x_PrivateSummary.exe?DoFn=.

9. more than $20 million: Ibid.

10. have now banned all travel: David D. Kirkpatrick, "Senate Passes Vast Overhaul in Ethics Rules," *New York Times,* January 19, 2007, http://select.nytimes.com/search/restricted/article?res=FA0F14FD35540C7A8DDDA80894DF404482.

11. Tammy Baldwin (D-WI) and her partner: "U.S. House of Representatives Travel Disclosure Form for Tammy Baldwin," PoliticalMoneyLine from *Congressional Quarterly,* February 14, 2006, http://www.politicalmoneyline.com/ptImg/2005/2/00000/00000544.pdf.

12. "government business dialogue": Travel Filings Statement by Robert Aderholt, Political MoneyLine from *Congressional Quarterly,* January 2006.

13. 1 After his trip to Eastern Europe, McDermott traveled to Santa Barbara, California, from January 20–22 to attend a conference on "Faith and Politics: A time for spiritual self reflection and open and honest dialogue." That trip was sponsored by the John E. Fetzer Institute, Inc., for a total cost of $9,794.

14. Rogers has received $76,000: "Campaign Finance Info: Rogers, Harold D," PoliticalMoney Line from *Congressional Quarterly,* http://www.politicalmoneyline.com/cgi-win/x_candpg.exe?DoFn=H0KY05015*2006.

15. Rogers and his wife: "Rogers, Harold Dallas—Gifts of Private Travel," PoliticalMoney Line from *Congressional Quarterly,* http://www.politicalmoneyline.com/cgi-win/x_Private Member.exe?DoFn=H0KY05015.

16. Twenty-six of those: Ibid.

17. "represented drug companies:" Josephine Hearn, "Records: Members Traveled on Lobbyists' Dime," *The Hill,* May 4, 2005, http://www.hillnews.com/thehill/export/TheHill/News/Frontpage/050405/records.html.

18. Accompanying Rogers on: Ibid.
19. Shaw received $120,600: PoliticalMoneyLine from *Congressional Quarterly*.
20. "Ashford Castle has": Hotel Description, "Enjoy Ireland's Premier Ashford Castle Hotel," http://www.irish-manors.com/ashford.html.
21. "The group spent": Hearn, "Records: Members Traveled on."
22. "Ripon Educational Fund": Ibid.
23. "While the Fund's": Karen Robb, "Richard Kessler and the Ripon Groups: How Lobbyists Give Lawmakers Free Trips Despite the Ban on Lobbyist-Funded Travel," *Public Citizen*, January 2006, http://www.citizen.org/documents/ripon.pdf.
24. paid almost $4 million: "Aspen Institute has funded these trips for Members of Congress," PoliticalMoneyLine from *Congressional Quarterly*, http://www.politicalmoneyline.com/cgi-win/x_PrivateSponsor.exe?DoFn=1987513.
25. "I could never," Eileen McGann interview with former senator Dick Clark (D-IA), August 30, 2006.
26. According to Clark: Ibid.
27. "the brightest member": Ibid.
28. she was paid $2,400 a year: David McCullough, *Truman* (New York: Simon and Shuster, 1992), 284.
29. "only just drop in": Ibid.
30. "no more than": "Federal Election Commission Advisory Opinion Number 2001-10," *Federal Election Comminssion*, July 17, 2001, http://herndon1.sdrdc.com/ao/no/010010.html.
31. "personal use of": Ibid.
32. she pled guilty: "Dana Tyrone Rohrabacher," SFGate.com, http://www.sfgate.com/cgi-bin/article.cgi?file=/politics/election/2002nov/bios/hca46rep.dtl.
33. "strategic guidance": Richard Simon, Chris Neubauer, and Rone Tempest, "Political Payrolls Include Families," *Los Angeles Times*, April 14, 2006, http://www.latimes.com/news/nation world/nation/la-na-campaign14apr14,0,2363055.story?coll=la-home-headlines.
34. at about $115,000: R. Jeffrey Smith, "Retirement Account of DeLay's Wife Traced," *Washington Post*, June 7, 2006, http://www.washingtonpost.com/wp-dyn/content/article/2006/06/06/AR2006060601320.html.
35. his daughter's Texas business –$350,304: Ibid.
36. "a recipe for Apple": Simon, Neubauer, and Rone, "Tempest, "Political Payrolls."
37. was paid $85,275: Ibid.
38. "sometimes catered the": Ibid.
39. "In the 2003–04 campaign": Ibid.
40. paid her husband's: Ibid.
41. paid his brother: Ibid.
42. paid his wife: David Gram, "Sanders campaign paid family members," Associated Press, April 13, 2005, http://www.boston.com/news/local/vermont/articles/2005/04/13/sanders_campaign_paid_family_members/.
43. a 15 percent commission: Ibid.
44. is firm Greenberg Traurig: Susan Schmidt and James V. Grimaldi, "Lawmakers Under Scrutiny in Probe of Lobbyist," *Washington Post*, November 26, 2005, http://www.washingtonpost.com/wp-dyn/content/article/2005/11/25/AR2005112501423.html.

45. Doolittle's recent FEC filing: Susan Crabtree and Jackie Kunich, "Doolittle Owes $139,000 to Wofe' Company," February 4, 2007.

46. paid her lobbyist son: Simon, Neubauer, and Tempest, "Political Payrolls Include."

47. paid his daughter-in-law: Ibid.

48. hired his wife: Ibid.

49. paid his wife: Ibid.

50. Lewis's stepdaughter: Charles R. Babcock and Alice R. Crites, "Congressman's Kin Got Pay from Lobbyists: Pac-Rep. Jerry Lewis' Ex-aide Is Biggest Contributor," *Washington Post,* June 8, 2006, http://www.sfgate.com/cgi-bin/article.cgi?file=/chronicle/archive/2006/06/08/MNGQIJAJM31.DTL&type=politics

51. received over $1 million: Ibid.

52. hired his wife: "DeLay Small Fish in Big Pond of Congressional Nepotism," Associated Press, April 14, 2005, http://www.newsmax.com/archives/articles/2005/4/13/212319.shtml

53. Wife Laurie Stupak earned: Ibid.

54. son Matthew received: Ibid.

55. hired his cousin: Ibid.

56. she received $114,804: Simon, Neubauer, and Tempest, "Political Payrolls Include."

57. to a baby shower: Susan Crabtree, "Abramoff, The Triplets and the Nanny Fund," *The Hill,* January 25, 2006, http://www.hillnews.com/thehill/export/TheHill/News/Frontpage/012506/news2.html.

58. any gift over $50: Ibid.

59. Wife Elizabeth was paid: "Lawmakers with Relatives on Payroll," Associated Press, April 13, 2005, http://www.sfgate.com/cgi-bin/article.cgi?f=/n/a/2005/04/13/national/w124553D99.DTL

60. hired his nephew: Ibid.

61. hired all three: Ibid.

62. hired his sister-in-law: Ibid.

63. employs his wife: Ibid.

64. hired his daughter Molly: Ibid.

65. paid his wife $505,000: Catlin Rother, "Lawmaker Keeps Wife on Payroll," SignOn SanDiego.com, December 4, 2005, http://www.signonsandiego.com/uniontrib/20051204/news_1m4filner.html.

66. pays his wife, Mary: Larry Margisak and Sharon Theimer, "Many Lawmakers Hire Family Members," *Desertnews.com,* April 14, 2005, http://deseretnews.com/dn/view/0,1249,600126073,00.html.

67. "financial independence": Ibid.

68. paid his wife, Laurie: Richard Simon, Chuck Neubauer, and Rome Tempest, "Political Payrolls Include Families," L.A. Times.com, April 14, 2006, *http://www.latimes.com/news/nation world/politics/la-na-campaign14apr14,1,7127738,full.story?coll=la-news-politics-national*

69. paid his wife, Gayle Ford: "Rep. John Sweeney (R-NY)," BeyondDelay.org, http://www.beyonddelay.org/summaries/sweeney.php.

70. paid him a $20,000 fee: John Kamman, "Campaign Committee Nepotism Under Fire," *Arizona Republic,* April 10, 2005, http://www.azcentral.com/arizonarepublic/news/articles/0410nepotism10.html.

71. "to fill in for": Ibid.

72. "installing campaign signs": Ibid.

73. "Neither federal law": Juliet Eilperin, "The Ties That Bind on Capitol Hill," *Washington Post,* August 7, 2003, http://www.washingtonpost.com/ac2/wp-dyn/A26010-2003Aug6?language=printer.

74. "in a class": Chuck Neubauer and Richard T. Cooper, "In Nevada, the Name to Know Is Reid," *Los Angeles Times,* June 23, 2003, http://www.westlx.org/HarryReid.pdf.

75. "has had three": Editorial, "Kith & Kin Inc.," *New York Times,* February 14, 2004, http://www.nytimes.com/2004/02/14/opinion/14SAT3.html?ex=1167282000&en=b411d6a235de17b3&ei=5070.

76. "A second son": Neubauer and Cooper, "In Nevada, the Name."

77. "In an internal": Ibid.

78. "Reid said . . . that": Ibid.

79. "while Reid has declared,": Robert D. Novak, "Reid and Reform," January 15, 2007, Townhall.com, http://www.townhall.com/columnists/RobertDNovak/2007/01/15/reid_and_reform.

80. "over $150,000 in": "Harry Reid [D-NV] The Most Corrupt Senator In DC," PipeLine News.com, http://www.pipelinenews.org/index.cfm?page=reidfixer.htm.

81. John Murtha (D-PA)'s brother: "Rep. John Murtha," BeyondDelay.org, http://www.beyonddelay.org/files/Murtha.pdf.

82. "Seven Dead Arson": Lynn Sweet, "Against History," Illinois Issues Online, September 2002, http://illinoisissues.uis.edu/features/2002sept/hastert.html.

83. "I realized that": Carl Hulse, "In Capitol, Last Names Link Some Leaders and Lobbyists," *New York Times,* August 4, 2002, http://www.commondreams.org/headlines02/0804-03.htm.

84. "In addition to": Biography, "Josh Hastert," PodestaMattoon.com, http://www.podestamattoon.com/biopages/JoshHastert.htm

85. the son represented Chiron: "Hastert, Joshua: Lobbyist Profile, 2005," OpenSecrets.org, http://www.opensecrets.org/lobbyists/lobbyist.asp?txtname=Hastert%2C+Joshua&year=2005.

86. 1 donated over $65,0000: OpenSecrets.com

87. Altria/Philip Morris profits: Jim VandeHei, "GOP Whip Quietly Tried to Aid Big Donor," *Washington Post,* June 11, 2003, http://www.washingtonpost.com/ac2/wp-dyn/A41839-2003Jun10?language=printer.

88. "thirty-one of those": "Rep. Roy Blunt: Ties to Special Interests Leave Him Unfit to Lead," Public Citizen's Congress Watch, January 2006, http://69.63.136.213/documents/Blunt_Final_Exec.pdf.

89. more than $270,000: James Ridgeway, "Sticky Fingers," *Village Voice,* January 17, 2006.

90. the Congressman's son: Thomas B. Edsall, "House Majority Whip Exerts Influence by Way of K Street," *Washington Post,* May 17, 2005, http://www.washingtonpost.com/wp-dyn/content/article/2005/05/16/AR2005051601334_pf.html.

91. "$1.5 million from": Hulse, "In Capitol, Last Names Link."

92. "When clients retain me": "Kith & Kin Inc.," *Congresspedia,* http://www.sourcewatch.org/index.php?title=Kith_%26_Kin_Inc.

93. work for Global Crossings: Wes Vernon, "Global Crossing Tied to Clinton Defense Secre-

tary," Newsmax.com, February 16, 2002, http://www.newsmax.com/archives/articles/2002/2/15/154416.shtml.

94. lobbyist for United Technologies: Jock Friedly, "For Hill spouses, lobbying begins at home," *The Hill,* September 17, 1997, http://www.alternet.org/columnists/story/4612/.

95. $137,000 from the: Chuck Neubauer, Judy Pasternak, and Richard T. Cooper, "Senator, His Son Get Boosts from Makers of Ephedra": *Los Angeles Times,* March 5, 2003, http://bioweb.usc.edu/courses/2003-spring/documents/bisc150-ca2_article10.pdf.

96. lobbyist for AT&T: Friedly, "For Hill spouses."

97. His wife, Catherine Stevens: "Kith & Kin Inc.," *Congresspedia.*

98. George F. Voinovich is a lobbyist: Ibid.

99. war criminal Slobodan Milosevic: "Add Karen Weldon, daughter of Rep. Curt Weldon, to the list of congressional sons, daughters, spouses, and in-laws whose lack of credentials, experience, or expertise hasn't stopped them from enjoying lucrative careers as government lobbyists," *Washington Monthly,* April 2004, http://www.findarticles.com/p/articles/mi_m1316/is_4_36/ai_n6006936.

100. "Family Members Lobbying Congress," Public Citizen, http://www.citizen.org/documents/Family%20Members%20Lobbying2.pdf.

101. Mrs. Baucas "pummeled,": "The Power of Being a U.S. Senator's Wife," CranialCavity.com, May 22, 2004, http://cranialcavity.net/wordpress/index.php/2004/05/22/the-power-of-being-a-us-senators-wife/.

102. Mrs. Baucus then: "Senator's Wife Charged in Dispute over Mulch," NBC4.com, April 22, 2004, http://www.nbc4.com/news/3028550/detail.html.

103. her "conflicting" stories: Richard Leiby, "The Reliable Source," *Washington Post,* April 22, 2004.

104. "swollen left cheek": "Senator's Wife Charged," www.NBC4.com.

105. "There was a situation": Ted Monsoon, "Mrs. Baucus Charged with Assault in D.C; Verbal Dispute Turned Physical Police Say," BillingsGazette.com, April 22, 2004, http://www.billingsgazette.com/newdex.php?display=rednews/2004/04/22/build/state/35-baucus-assault.inc.

106. "the truth will come out": Richard Leiby, "The Reliable Source," *Washington Post,* May 5, 2004.

107. "oddly carefree": Betsy Rothstein, "Wives Like Wanda," *The Hill,* May 11, 2004, http://www.hillnews.com/living/051104_wives.aspx.

108. "Get away from there": Richard Leiby, "The Reliable Source," *Washington Post,* May 5, 2004.

109. "She was very sweet": Beppe Severgnini, *Ciao, America!: An Italian Discovers the U.S.* (New York: Broadway, 2002).

110. "camels that are": Richard Leiby, "The Reliable Source," *Washington Post,* March 26, 2003.

111. "A billion Muslims": Ibid.

112. projects costing $28 billion: Jodi Rudoren and Aron Pilhofer, *New York Times,* July 2, 2006, http://www.nytimes.com/2006/07/02/washington/02earmarks.html?ex=1309492800&en=c4371dd2dab7497c&ei=5088&partner=rssnyt&emc=rss.

113. "The town of Ketchikan": John Stossel, Glenn Ruppel, and Ann Varney, "The Real Price of Pork Barrel Spending," ABC News, Sept. 9, 2005, http://abcnews.go.com/2020/GiveMeABreak/story?id=1112408&page=1.

114. Robert D. Novak, "Harry Reid vs. Reform," *New York Post,* January 15, 2007, 23..

115. "Frank, you ought to": Ted Morgan, *FDR* (New York: Simon and Schuster, 1985),123.

116. an amendment giving $800,000: Michael McAuliff, "Hillary Decks Hall of Fame with 800G," New York *Daily News,* July 21, 2006, http://www.nydailynews.com/news/regional/story/ 436861p-368098c.html.

117. future House Leader Tom Delay: Ken Silverstein, "The Great American Pork Barrel: Washington Streamlines the Means of Corruption," Harpers.org of *Harper's* magazine, July 2005. Sources

118. The Office of Naval Research: Charles R. Babcock, "The Project That Wouldn't Die: Using earmarks, members of Congress kept money flowing to a local company that got $37 million for technology the military couldn't use," *Washington Post,* June 19, 2006 page D01 at http:// www.washingtonpost.com/wp-dyn/content/article/2006/06/18/AR2006061800631.html.

119. "conceived as a way": Ibid.

120. "It kept failing": Ibid.

121. $17,000 in campaign contributions: Ibid.

122. Rajant got its contract: Silverstein, "The Great American Pork."

123. "This disaster": Stossel, Ruppel and Varney, "The Real Price of Pork."

124. "a chilling preview": Michael Grunwald and Susan B. Glasser, "The Slow Drowning of New Orleans," *Washington Post,* October 9, 2005, http://www.washingtonpost.com/wp-dyn/ content/article/2005/10/08/AR2005100801458.html.

125. "before Hurricane Katrina": Michael Grunwald, "Largess in Louisiana: Money Flowed to Questionable Projects," *Washington Post,* September 8, 2005, http://www.washingtonpost .com/wp-dyn/content/article/2005/09/07/AR2005090702385.html.

126. "Louisiana has received": Ibid.

127. "flunked a Corps cost-benefit": Ibid.

128. "also spends tens": Ibid.

129. "in 1998, the Corps": Ibid.

130. "we all should have paid": Ibid.

131. "they could have built": Ibid.

132. for the "Prairie Parkway": Jonathan Weisman, "Lawmakers' Profits Are Scrutinized: Hastert and Others Defend Land Gains," *Washington Post,* June 22, 2006, http://www.washington post.com/wp-dyn/content/article/2006/06/21/AR2006062102210.html.

133. "a $2 million profit": Ibid.

134. "previously hemmed-in land": Ibid.

135. TK: Ibid.

136. "honest graft": William Riordan, *Plunkett of Tammany Hall* (New York: Dutton, 1963), http://www.marxists.org/reference/archive/plunkett-george/tammany-hall/index.htm#s01.

137 "$2.3 million for the": Jonathan Weisman and Jim VandeHei, "Road Bill Reflects the Power Of Pork," *Washington Post,* August 11, 2005, http://www.washingtonpost.com/wp-dyn/ content/article/2005/08/10/AR2005081000223_pf.html.

138. "virtually all of its income": Silverstein, "The Great American Pork Barrel." Sources

139. $1 million Livingston got: Ibid.

140. "slash science": John Kelly, "Pork threatens NASA plans: Congress' Pet Projects Take $3 Billion from Budget," *Florida Today/USA Today,* June 12, 2006.

141. under investigation: Jodi Rudoren, "Lawmaker and Beneficiary Bought Farm," *New York Times,* April 26, 2006.

142. "Mr. Mollohan said": Ibid.

143. got almost $30 million: Ibid.

144. "no evidence that Mr. Mollohan": Jodi Rudoren and Aron Pilhofer, "Congressman's Condo Deal Is Examined," *New York Times,* May 17, 2006.

145. $1.2 million in income: Jodi Rudoren, David Johnston, and Aron Pilhofer, "Special Projects by Congressman Draw Complaints," *New York Times,* April 8, 2006.

146. an estimated $480 million: Ibid.

147. "from 1997 through February 2006": Rudoren, Johnston, and Pilhofer, "Special Projects Draw Complaints."

148. More than 3,000 companies: Silverstein, "The Great American Pork." Sources

149. $103.1 million to members of Congress : Report, "Lobbyists Contributed $103 Million to Lawmakers Since 1998," Public Citizen, May 22, 2006, http://www.citizen.org/pressroom/release.cfm?ID=2203.

150. Report, "The Bankrollers: Lobbyists' Payments to the Lawmakers 1998–2006," Public Citizen, May 2006, http://www.citizen.org/documents/BankrollersFinal.pdf.

151. Delay got $1,322,906: Ibid.

152. Members of the House Who Received at Least $500,000 from Lobbyists: Ibid.

153. A total of $640 million: Jodi Rudoren and Aron Pilhofer, "Hiring Lobbyists for Federal Aid, Towns Learn That Money Talks," *New York Times,* July 2, 2006, http://www.nytimes.com/2006/07/02/washington/02earmarks.html?ex=1309492800&en=c4371dd2dab7497c&ei=5088&partner=rssnyt&emc=rss.

154. Now 1,421 hire lobbyists: Ibid.

155. "Today the halls": "When Pork-Barrel Pols Aren't Enough," Business Week Online, June 12, 2006, http://www.businessweek.com/magazine/content/06_24/b3988080.htm.

156. "We need to be there": Ibid.

157. "worth every penny": Rudoren and Pilhofer, "Hiring Lobbyists for Federal Aid."

158. "to significantly participate": Ibid.

159. "It goes beyond mere": Ibid.

160. Republican Senator Rick Santorum: "Report: Santorum, Clinton Took Most from Lobbyists," Rawstory.com, May 23, 2006, http://www.rawstory.com/news/2006/Report_Santorum_Clinton_took_most_from_0523.html.

161. "In April 2003": Mike Mcintire and Raymond Hernandez, "Company Finds Clinton Useful, and Vice Versa," *New York Times,* April 12, 2006, http://www.nytimes.com/2006/04/12/nyregion/12hillary.html?ex=1302494400&en=224ffa23dd8611c9&ei=5088&partner=rssnyt&emc=rss.

162. "in April 2004": Ibid.

163. $137,000 in contributions: Mike Mcintire and Raymond Hernandez, "Company Finds Clinton Useful, and Vice Versa," *New York Times,* April 12, 2006, http://www.nytimes.com/2006/04/12/nyregion/12hillary.html?ex=1302494400&en=224ffa23dd8611c9&ei=5088&partner=rssnyt&emc=rss.

164. Stevens has gotten $325 million: "2006 Congressional Pig Book -Annual Compilation of the Pork-Barrel Projects in the Federal Budget," Citizens Against Government Waste, http://www.cagw.org/site/PageServer?pagename=reports_pigbook2006.

165. $25,000,000 for rural: Ibid.

166. Cochran's pet earmarks include: Ibid.

167. $1.5 million for the: Robert D. Novak, "Log-rolling for pork," Townhall.com, Monday, June 19, 2006, http://www.townhall.com/columnists/RobertDNovak/2006/06/19/log-rolling_for_pork.

168. "We say that": Ibid.

169. "every President from": "Cured Pork—Remedies for Earmarking and Other Congressional Addictions," *Wall Street Journal,* January 28, 2006, http://www.opinionjournal.com/week end/hottopic/?id=110007887.

170. up by about $31,000: Monisha Bansal, "Dems Willing to Sacrifice Pay Hikes for Minimum Wage Increases," CNSnews.com, July 13, 2006, http://www.cnsnews.com/ViewPolitics.asp?Page=/Politics/archive/200607/POL20060713a.html.

171. about $10,700 a year: Kathy Kiely, "Some Lawmakers Balk at Proposed Salary Increases," *USA Today,* June 25, 2006, http://www.usatoday.com/news/washington/2006-06-25-congress-pay_x.htm.

172. an extra $3,300: "Your congressman Just Got A Raise," CNN.com, July 13, 2006, http://www.cnn.com/2006/POLITICS/06/13/congress.payraise.ap/index.html.

173. were paid $6.00 a day: "Senate Salaries Since 1789," United States Senate, http://www.senate.gov/artandhistory/history/common/briefing/senate_salaries.htm.

174. "tho we may set": the U.S. Constitution, "Article 2, Section 1, Clause 7," University of Chicago Press, http://press-pubs.uchicago.edu/founders/documents/a2_1_7s2.html.

175. "Senate Salaries since 1789": "Senate Salaries."

CHAPTER FIVE

1. "Never . . . was so much owed": Sir Winston Churchill, speech to the House of Commons of the British Parliament, August 20, 1940, http://en.wikipedia.org/wiki/So_much_owed_by_so_many_to_so_few.

2. 144,000 cars cross: "Brooklyn Bridge History," aamcar.com, http://www.aamcar.com/brooklyn-bridge.php.

3. planning the 9/11 attacks: Eric Lichtblau with Monica Davey, "Threats and Responses: Terror; Suspect in Plot on Bridge Drew Interest Earlier," *New York Times,* June 21, 2003, http://select.nytimes.com/search/restricted/article?res=F70711F83D5F0C728EDDAF0894 DB404482.

4. "sites for possible terrorist attacks: "Alleged Terror Trucker Pulls Plea," CBS/AP, September 26. 2003, http://www.cbsnews.com/stories/2003/06/20/terror/main559532.shtml.

5. "tickets and cell phones": Ibid.

6. "directly to cargo planes": Ibid.

7. "wife, Geneva Bowling": Lichtblau with Davey, "Threats and Responses."

8. "drawn to Mr. Faris": Ibid.

9. "one issue of interest": Ibid.

10. "A transcript of Faris's": Mark Hosenball, "Did It Work?" *Newsweek,* January 4, 2006, http://www.msnbc.msn.com/id/10711930/site/newsweek/.

11. "overseas source information": Ibid.

12. officials had claimed: Ibid.

13. "the bridge in the": Authors' interview with Commissioner Ray Kelly, November 20, 2006.

14. bridge was "too hot": Ibid.

15. 1 Faris confessed and was sentenced: CBS News, "Feds Allege Ohio Mall Terror Plot," June 14, 2004, http://www.cbsnews.com/stories/2004/06/15/terror/main623119.shtml.

16. "The American heartland": Ibid.

17. "anti-American views": Hosenball, "Did It Work?"

18. "It is unclear": Ibid.

19. "A lengthy statement Babar": Ibid.

20. "by using controversial provisions": Ibid.

21. 1 "provision might permit": Senator Russell Feingold, "On Opposing the U.S. Patriot Act," October 12, 2001, http://www.archipelago.org/vol6-2/feingold.htm.

22. "having documented connections": Michele Morgan Bolton, "Sting Case Prosecutors Ask to Shield Witnesses: Anonymity for Translators Will Prejudice Jury, Argue Attorneys for Mosque Leader, Pizza Shop Owner," *Albany Times Union*, August 23, 2006.

23. "intelligence for Osama bin Laden": Ibid.

24. made contact with Abu Zubaydah: "Profile: Jose Padilla," BBC, November 22, 2005, http://news.bbc.co.uk/2/hi/americas/2037444.stm.

25. "threats against the nation": Richard B. Schmitt, "7 Arrested in Miami in Alleged Terrorist Plot: They reportedly talked of attacking Chicago's Sears Tower and a Florida FBI office," *Los Angeles Times*, June 23, 2006 http://www.latimes.com/news/nationworld/nation/la-na-plot23jun23,0,7916402.story?coll=la-home-headlines.

26. "FBI office in Miami": Ibid.

27. "nimble as possible": "Arrests Cast New Light on British Anti-terror Policies," PBS.org, August 15, 2006, http://www.pbs.org/newshour/bb/terrorism/july-dec06/terrorism_08-15 .html.

28. "both because of": Ibid.

29. "[It's] a tremendously important:" Ibid.

30. "Once you are formally": Ibid.

31. "there is a difference": Ibid.

32. "In the United States": Ibid.

33. a "congestion tax": "London congestion charge," *Wikipedia*, http://en.wikipedia.org/wiki/ London_congestion_charge.

34. "video records are": PBS.org, "Arrests Cast New Light."

CHAPTER SIX

1. "You have Gulliver": Howard Fuller and George Mitchell, "A Culture of Complaint," *Hoover Institution/EducationNext*, 2006, http://www.hoover.org/publications/ednext/3211 456.html.

2. "adjusted for inflation" "Per Pupil Education Spending In Constant Dollars": U.S. Department of Education, National Center for Education Statistics: Common Core of Data (CCD), "National Public Education Financial Survey," 1988–89 through 2001–02; National Elementary and Secondary Average Daily Attendance Model, 1991–92 through 2001–02; and Elementary and Secondary School Current Expenditures Model, 1969–70 through 2001–02, prepared October 2004, "Table 34: Actual and alternative projected numbers for current expenditures and current expenditures per pupil in average daily attendance in public ele-

mentary and secondary schools: 1988–89 to 2013–14," from: http://nces.ed.gov/programs/projections/TableDisplay.asp?id=tab_34.asp&a=elemsec.

3. "student achievement scores": U.S. Department of Education, Institute of Education Sciences, National Center for Education Statistics, National Assessment of Educational Progress (NAEP), 1992–2005 Reading Assessments.

4. "and math scores": Ibid.

5. a study by the Hoover Institution: Eva Moskowitz, "Breakdown," Hoover Institution/EducationNext, 2006, http://www.hoover.org/publications/ednext/3211506.html.

6. Joel Klein reports that: John Stossel, "Stupid in America: How Lack of Choice Cheats Our Kids Out of a Good Education," ABC News, January 13, 2006, http://abcnews.go.com/2020/Stossel/story?id=1500338 .

7. "It took years": Ibid.

8. "klein said he": Ibid.

9. "I believe that leaders": Ibid.

10. "the chancellor's office": David M. Herszenhorn, "Schools Chief Says Union Rules Lead Him to Create Unneeded Jobs," New York Times, August 31, 2006, http://www.nytimes.com/2006/08/31/nyregion/31schools.html?ex=1314676800&en=ab2fb70b61909555&ei=5090&partner=rssuserland&emc=rss.

11. "If these assistant principals": Ibid.

12. cost to the City of $5 million: Ibid.

13. 1 In Milwaukee, for example: Fuller and Mitchell, "A Culture of Complaint."

14. "the annual migration": Mercury News Editorial, "State needs 'lemon' law for teachers—Bill Before Governor Allows Schools To Compete For The Best And Avoid The Worst," San Jose Mercury News, September 1, 2006, http://nl.newsbank.com/nl-search/we/Archives?s_site=mercurynews&p_multi=SJ|&p_product=SJ&p_theme=realcities&p_action=search&p_maxdocs=200&p_text_search-0=state%20AND%20needs%20AND%20lemon%20AND%20law&s_dispstring=state%20needs%20lemon%20law%20AND%20date(last%20180%20days)&p_field_date-0=YMD_date&p_params_date-0=date:B,E&p_text_date-0=-180qzD&xcal_numdocs=20&p_perpage=10&p_sort=YMD_date:D&xcal_useweights=no.

15. "Per Pupil Education Spending In Constant Dollars": U.S. Department of Education, National Center for Education Statistics: Common Core of Data (CCD), "National Public Education Financial Survey," 1988–89 through 2001–02; National Elementary and Secondary Average Daily Attendance Model, 1991–92 through 2001–02; and Elementary and Secondary School Current Expenditures Model, 1969–70 through 2001–02, prepared October 2004, "Table 34: Actual and alternative projected numbers for current expenditures and current expenditures per pupil in average daily attendance in public elementary and secondary schools: 1988–89 to 2013–14," from: http://nces.ed.gov/programs/projections/TableDisplay.asp?id=tab_34.asp&a=elemsec.

16. "On the whole": "Helping All Students Learn: Identifying School Districts Across the U.S. that are Significantly Narrowing Achievement Gaps," Standard & Poor's, http://www.schoolmatters.com/app/content/q/mtype=schoolmatters_National_achievement_gap.shtml/mlvl=1/stid=1036196/llid=162/stllid=676/locid=1036195/site=pes.

17. "in the majority": Ibid.

18. "only 2 percent": Lowell C. Rose and Alec M. Gallup, "The 36th Annual Phi Delta Kappa/

Gallup Poll of the Public's Attitudes Toward the Public Schools," PDK International, August 19, 2004, http://www.pdkintl.org/kappan/k0409pol.htm.

19. Among parents: Ibid.

20. How Parents Rate The Schools: Rose and Gallup, "The 36th Annual Phi Delta."

21. How Adults with No Children in School Rate Schools: Ibid.

22. 381 to 41 in the House: "No Child Left Behind Act," *Wikipedia,* http://en.wikipedia.org/wiki/No_Child_Left_Behind_Act.

23. "parents with children": "Fact Sheet: No Child Left Behind Act," The White House, http://www.whitehouse.gov/news/releases/2002/01/20020108.html.

24. Most often, charter schools: Report, "Charter Schools Today: Changing the Face of American Education—Part 1: Annual Survey of America's Charter Schools, 2005 Data," Center For Education Reform, http://www.edreform.com/_upload/cer_charter_survey.pdf.

25. "who now runs": Stossel, "Stupid in America."

26. "To save money": Ibid.

27. "his school also": Ibid.

28. "It's not about": Ibid.

29. "Give me the poor": Ibid.

30. 1 million students enrolled: Report, "Charter Schools Today."

31. enroll an average of 758: Ibid.

32. 13 percent have both: Ibid.

33. now spends 70 percent: "Pride and Promise of Iowa Public Schools," http://www.myiowaschools.org/iowa_school_pride_facts.asp.

34. public schools average $8800: Report, "Charter Schools Today."

35. and Texas (85,000): Ibid.

36. 17 percent better: Ibid.

37. 12 percent more proficient: Ibid.

38. 13 percent more likely: Ibid.

39. 21 percent among blacks statewide: Ibid.

40. collects $1 billion: Hans Zeiger, "NEA vs. America's Future," *World Net Daily,* July 23, 2004, http://www.worldnetdaily.com/news/article.asp?ARTICLE_ID=39598.

41. "When school children: Patrick Chisholm, "The Misnamed National Education Association," *Christian Science Monitor,* August 24, 2005, http://www.csmonitor.com/2005/0824/p25s01-cogn.html.

42. "peer review of teachers": Julia E. Koppich, "A Tale of Two Approaches—The AFT, the NEA, and NCLB," *Peabody Journal of Education* 80, no. 2 (2005): 137–55, http://www.leaonline.com/doi/abs/10.1207/S15327930pje8002_8.

43. "From the beginning": Ibid.

44. "a one-size-fits-all": Koppich, "A Tale of Two Approaches."

45. "The federal NCLB Act": Ibid.

46. NCLB law are: "CTA Statewide Ad Campaign and NEA Data Tell How Federal Education Law Leaves Millions of Kids Behind," January 16, 2004, http://www.cta.org/media/newsroom/releases/archive/2004/20040116PR.htm

47. not sufficiently funded: Koppich, "A Tale of Two Approaches."

48. "takes testing to extremes": Ibid.

49. six million children: News Release, "Back to School: 2006–2007," August 16, 2006, http://www.census.gov/Press-Release/www/releases/archives/facts_for_features_special_editions/007108.html.

50. "due process when": Koppich, "A Tale of Two Approaches."

51. Arkansas' innovative former: Mike Huckabee, *From Hope to Higher Ground* (New York: Center Street, 2006), 45.

52. "veteran teachers are": Koppich, "A Tale of Two Approaches."

53. "spent $1,871,000 on": PoliticalMoneyLine from *Congressional Quarterly,* http://www.fecinfo.com/cgi-win/indexhtml.exe?MBF=NAME.

54. rose by 26 percent : "Estimated average annual salary of teachers in public elementary and secondary schools: Selected years, 1959–60 to 2002–03," *National Center For Educational Statistics,* http://nces.ed.gov/programs/digest/d04/tables/dt04_077.asp.

55. "is $54,689, about 15 percent": Annual Demographic Survey: March Supplement," U.S. Census Bureau, http://pubdb3.census.gov/macro/032006/perinc/new04_001.htm.

CHAPTER SEVEN

1. "Canada spends $2,222": "The World Health Report 2005," World Health Organization.

2. "$250.6 billion a year": Matthew Herper and Peter Kang, "The World's 10 Best-Selling Drugs," Forbes.com, March, 28, 2006, http://www.forbes.com/home/sciencesand medicine/2006/03/21/pfizer-merck-amgen-cx_mh_pk_0321topdrugs.html.

3. 1 $800 million: Christopher Rowland, "Drug Ads Deliver a Few Side Effects," *Boston Globe,* June 12, 2003.

4. $4.2 billion in 2005: Report Summary, "Prescription Drugs: Improvements Needed in FDA's Oversight of Direct-to-Consumer Advertising," U.S. Government Accountability Office, November 16, 2006, http://www.gao.gov/docdblite/details.php?rptno=GAO-07-54.

5. every 4.7 doctors: Henry A. Waxman, J.D. "The Lessons of Vioxx—Drug Safety and Sales," *New England Journal of Medicine,* June 23, 2005, http://content.nejm.org/cgi/content/full/352/25/2576?query=TOC.

6. "in 1997 when": Alexandra Marks, "Rise of 'Ask Your Doctor' Ads: A Public-Health Concern?" *Christian Science Monitor,* November 30, 2001, http://www.csmonitor.com/2001/1130/p2s2-ussc.html.

7. "Almost 6 percent": "100 Leading National Advertisers: Profiles of the Top 100 U.S. Marketers in This 51st Annual Ranking," *Advertising Age,* June 26, 2006, http://adage.com/images/random/LNA2006.pdf.

8. "doing something that": Alex Berenson, "Indictment of Doctor Tests Drug Marketing Rules," *New York Times,* July 22, 2006, http://www.nytimes.com/2006/07/22/business/22drugdoc.html?ex=1311220800&en=eeafd3a0752f4924&ei=5088&partner=rssnyt&emc=rss.

9. "in 2002, according to": Carl Elliott, "The Drug Pushers," *The Atlantic,* April 2006, http://www.theatlantic.com/doc/prem/200604/drug-reps.

10. "make a 16.2 percent profit ": Wendy Diller and Herman Saftlas, "Healthcare: Pharmaceuticals," *Standard & Poor's Industry Surveys,* December 22, 2005,13.

11. The Top Ten Drug Companies: Ibid.

12. "The way it": Bernadette Tansey, "Hard Sell: How Marketing Drives the Pharmaceutical Industry," *San Francisco Chronicle*, February 27, 2005, http://sfgate.com/cgi-bin/article.cgi?file=/c/a/2005/02/27/MNGUOBHO781.DTL.

13. more than $100 in value: Gardiner Harris, "As Doctors Write Prescriptions, Drug Company Writes a Check," *New York Times*, June 27, 2004, http://www.nytimes.com/2004/06/27/business/27DRUG.final.html?ex=1403668800&en=d74fd7b7eb933e39&ei=5007&partner=USERLAND.

14. participate in clinical trials: David Blumenthal, MD, MPP, "Doctors and Drug Companies," *New England Journal Of Medicine*, October 28, 2004, http://content.nejm.org/cgi/content/extract/351/18/1885.

15. "cross the line": Dick Morris interview with Dr. Robert Cowan, June 8, 2006.

16. a "consulting" agreement: Harris, "As Doctors Write Prescriptions."

17. "were little more than": Ibid.

18. $8.3 billion of sales in 2003: Ibid.

19. "was aggressively marketed": Department of Justice, Press Release, "Warner-Lambert to Pay $430 Million to Resolve Criminal & Civil Health Care Liability Relating to Off-Label Promotion," May 13, 2004, http://www.usdoj.gov/opa/pr/2004/May/04_civ_322.htm.

20. bill the government: Harris, "As Doctors Write Prescriptions."

21. "the pain killers Vioxx": Tansey, "Hard Sell: How Marketing."

22. $900 million in sales: Ibid.

23. "study . . . showed that": J. D. Waxman, "The Lessons of Vioxx."

24. "more than 100 million": Ibid.

25. "should be made": Ibid.

26. "the next day": Ibid.

27. "The Cardiovascular Card": Ibid.

28. "serious concerns": Ibid.

29. "studies that raised": Ibid.

30. "so friendly, so": Elliott, "The Drug Pushers."

31. "one rep told me": Ibid.

32. "an arms race": Ibid.

33. "The trick is to": Ibid.

34. drug companies provided: Ibid.

35. "nearly twice as": Ibid.

36. Dr. DeAngelis said: Dr. Catherine D. DeAngelis: "Journal Editor Again Says She Was Misled," Associated Press, July 19, 2006, http://query.nytimes.com/gst/fullpage.html?res=9403E3DA163FF93AA25754C0A9609C8B63.

37. 12 percent of the growth: Christopher Rowland, "Drug Ads Deliver a Few Side Effects," *Boston Globe*, June 12, 2003, http://www.dtcperspectives.com/content.asp?id=157.

38. "one purpose": Robert Pear, "Drug Industry Is Said to Work on an Ad Code," *New York Times*, May 17, 2005, http://www.nytimes.com/2005/05/17/politics/17drug.html?ex=1168059600&en=ae17deafeebf2ca2&ei=5070.

39. "Better to self-regulate": Sourcewatch, "Pharmaceutical Research and Manufacturers of America," Center for Media and Democracy, http://www.sourcewatch.org/index.php?title=Pharmaceutical_Research_and_Manufacturers_of_America.

40. tax money generates 39 percent: Cathy A. Cowan, MBA. and Micah B. Hartman, "Financing Health Care: Businesses, Households, And Government, 1987–2003," Department of HHS, http://www.cms.hhs.gov/HealthCareFinancingReview/downloads/Web_Exclusive_Cowan .pdf.

41. $150 million in contributions : Victoria Kreha, "Checkbook Politics: Over the last seven years, the pharmaceutical industry has given $150 million in campaign contributions," Center for Public Integrity, July 7, 2005, http://www.publicintegrity.org/rx/report.aspx?aid= 720.

42. Top Twenty Recipients of Drug Company Donations: Ibid.

43. "527 organizations": Ibid.

44. "a government bureaucrat": Robert Pear, "House Democrats Pass Bill on Medicare Drug Prices," New York Times, January 12, 2007, http://query.nytimes.com/gst/fullpage.html?sec= health&res=9F00E6DE1330F931A25752C0A9619C8B63.

45. "he will veto it": Ibid.

46. "It's a very bad precedent": Jordan Rau, "Industry Aims to Defeat Discount Drug Initiatives," Los Angeles Times, Mar 28, 2005, available at: http://www.sourcewatch.org/index.php? title=Pharmaceutical_Research_and_Manufacturers_of_America.

47. "voluntary discount plan": Ibid.

48. "defeated by 1.6 million": "State Ballot Measures, Statewide Returns for California Special Statewide Election," California Secretary of State website, November 8, 2005, available at: http://vote2005.ss.ca.gov/Returns/prop/00.htm.

49. "to infuse low-priced" "Joe Nocera, "Generic Drugs: The Window Has Loopholes," New York Times, July 1, 2006, http://select.nytimes.com/2006/07/01/business/01nocera.html.

50. "monopoly status: Ibid.

51. "53 percent of prescriptions": Marc Kaufman, "Drug Firms' Deals Allowing Exclusivity Makers of Generics Being Paid to Drop Patent Challenges, FTC Review Finds," Washington Post, April 25, 2006..

52. "an extraordinary time": Nocera, "Generic Drugs."

53. Schering-Plough and: Kaufman, "Drug Firms' Deals."

54. "carte blanche to": Ibid.

55. sleep aide, trazodone: Daniel Carlat, "Generic Smear Campaign," New York Times, May 9, 2006, http://www.nytimes.com/2006/05/09/opinion/09carlat.html?ex=1304827200&en=a4 c066300b3c203a&ei=5088&partner=rssnyt&emc=rss.

56. "is categorized as": Ibid.

57. "In the past few years": Ibid.

58. "like cardiac arrhythmias": Ibid.

59. "predatory pricing": Nocera, "Generic Drugs."

60. $4.4 billion in revenues: Ibid.

61. "largest-selling drug": Ibid.

62. "the new scam": Ibid.

CHAPTER EIGHT

1. Franklin Delano Raines: "Porker of the Month," Citizens Against Government Waste, May 2004, http://www.cagw.org/site/PageServer?pagename=news_porkerofthemonth_May04.

2. The Other Deserving Democrats: Ross Guberman, "Balancing Act: Fannie Mae Projects a Happy Image. But as Its Debt Grows Bigger and Its Executives Get Richer, Should Taxpayers Start to Worry?" Washingtonian.com, August 2002.

3. $25 million in salaries and bonuses: "Fannie Mae's Fall from Grace: Scandal Escapes Attention," *Rocky Mountain News*, August 16, 2006, http://www.rockymountainnews.com/drmn/editorials/article/0,2777,DRMN_23964_4911017,00.html.

4. And Some Republicans Too: Guberman, "Balancing Act."

5. "$1 million in total compensation": Bethany McLean, "The Fall of Fannie Mae," *Fortune* magazine, from CNN Money.com, January 24, 2005, available at http://money.cnn.com/magazines/fortune/fortune_archive/2005/01/24/8234040/index.htm.

6. "describe their bonds": Guberman, "Balancing Act."

7. "We have so many": Ibid.

8. "The subsidy to": Ibid.

9. "for every $3": Ibid.

10. profit of $6 billion: Ibid.

11. "Our mission is": Franklin D. Raines, "Fannie Mae 2004 Annual Shareholders' Meeting," FannieMae.com, May 25, 2004, http://www.fanniemae.com/media/speeches/speech.jhtml?repID=/media/speeches/2004/speech_244.xml&p=Media&s=Executive+Speeches&counter=2.

12. "Its time to say": Guberman, "Balancing Act."

13. coming in at $118,000: Ibid.

14. "for years HUD": Ibid

15. "Together, Fannie Mae": Ibid.

16. "the taxpayers are": Ibid.

17. "arguably the most": Ibid.

18. "the fact that Frank Raines": TV program, "Big-Hearted Companies,", *Wall $treet Week*, with *Fortune,* January 14, 2005, http://www.pbs.org/wsw/tvprogram/20050114.html.

19. "used charitable contributions": Bruce Meyerson, "Corporate Philanthropy, Though Popular, Is Also Open to Abuse," Associated Press, March 04, 2006, http://www.post-gazette.com/pg/06063/663755.stm.

20. "millions of dollars": Ibid.

21. "organizations whose board": Ibid.

22. half that projected rate: Ibid.

23. "Fannie has tried": Ibid.

24. "in a confidential report": Ibid.

25. "senior management manipulated": "Fannie Mae Board, Management cited for Misconduct," Forbes.com, May 23, 2006, http://www.forbes.com/markets/feeds/afx/2006/05/23/afx2768468.html.

26. "blasts [Raines] for his role": TV Program, "Fannie Mae Agrees to Pay," PBS.Org, May 23, 2006, http://www.pbs.org/nbr/site/onair/transcripts/060523g/.

27. "arrogant and unethical": Ibid.

28. "Rolodex hiring": Guberman, "Balancing Act."

29. "after the *Wall Street Journal*": Ibid.

30. "after sending around": Ibid.

31. "hired a phone bank": Ibid.

32. Fannie Mae Runs Ads: McLean, "The Fall of Fannie Mae."
33. "Here is an organization": McLean, "The Fall of Fannie Mae."
34. "Fannie sprinkles millions": Guberman, "Balancing Act."
35. "foundation had existed": McLean, "The Fall of Fannie Mae."
36. "free, step-by-step": "Fannie Mae Foundation," DiscoverTheNetwork.org, 2006, http://dis coverthenetworks.com/funderprofile.asp?fndid=5197&category=78.
37. Organizations That Got Grants: Ibid.
38. "Fannie Mae is": Guberman, "Balancing Act."
39. received letters opposing: Ibid.
40. "he picketed Fannie's": Ibid.
41. got fourteen checks: Ibid.
42. "partnership offices": Ibid.
43. "you did not question": McLean, "The Fall of Fannie Mae."
44. "in 1996, when": Ibid.
45. "called for a new": Ibid.
46. "disrupting the capital": Ibid.
47. "special examination of": Press Release, "OFHEO Report: Fannie Mae Facade: Fannie Mae Criticized For Earnings Manipulation," Office of Federal Housing Enterprise Oversight (OFHEO), May 23, 2006, http://www.ofheo.gov/media/pdf/fnmserelease.pdf.
48. "allowed the company": Ibid.
49. "these accounting standards": Ibid.
50. "This hearing is": Ibid.
51. "Is it possible": McLean, "The Fall of Frank Raines."
52. "overplayed his hand": Ibid.
53. "At a little after 6:00 p..m.": McLean, "The Fall of Fannie Mae."
54. "Raines looked stricken": Ibid.
55. "Fannie employees manipulated": Press Release, "OFHEO Report: Fannie Mae Facade."
56. implement thirteen recommendations: Ibid.
57. "To the extent": McLean, "The Fall of Fannie Mae."
58. "We have all": Press Release, "Fannie Mae Agrees to Comprehensive Settlements with OFHEO and SEC," FannieMae.com, May 23, 2005, http://www.fanniemae.com/news releases/2006/3738.jhtml?p=Media&s=News+Releases.
59. "Mudd's not out": Robert Schröder, "Fannie Mae cited for misconduct: Mortgage finance giant agrees to $400 million fine," MarketWatch.com, May 23, 2006, http://www.market watch.com/news/story/fannie-mae-board-management-cited/story.aspx?guid=%7B010C3 F92-94F9-4DDF-873D-9BC40810A439%7D.

CHAPTER NINE

1. $30 billion in 2004: Stephen Labaton, "Bankruptcy Bill Set for Passage; Victory for Bush," New York Times, March 9, 2005, http://www.nytimes.com/2005/03/09/business/09bank ruptcy.html?ex=1268110800&en=05c9987df2e25e4b&ei=5088.
2. Only 2 percent are: Jonathan Alter, "A Bankrupt Way to Do Business," Newsweek, April 25, 2005, http://www.msnbc.msn.com/id/7528519/site/newsweek/.
3. 1.6 million in 2004: Labaton, "Bankruptcy Bill Set."

4. one in sixty: Liz Pulliam Weston, "Bankruptcy filings soaring again," MSN.com, 2006, http://articles.moneycentral.msn.com/Banking/BankruptcyGuide/BankruptcyFilingsSoar ingAgain.aspx.

5. frequently exceed 30 percent: Ibid.

6. 45 million uninsured Americans: Ibid.

7. About 75 percent: R.V. Scheide, "You're Going Broke!" Newsreview.com, March 2, 2006, http://www.newsreview.com/reno/Content?oid=oid%3A47519.

8. $83,000 in Massachusetts: Labaton, "Bankruptcy Bill Set."

9. only 42-0.3 percent: Scheide, "You're Going Broke!"

10. $100,000 per day: Weston, "Bankruptcy filings soaring."

11. "a list of secured": Joan E. Lisante, "New Bankruptcy Law Tightens Rules, Adds Paperwork," ConsumerAffairs.Com, October 14, 2005, http://www.consumeraffairs.com/news04/2005/bankruptcy_2005.html.

12. "could have addressed": Editorial, "How a Bad Bill Becomes Law," San Francisco Chronicle, March 13, 2005, http://www.sfgate.com/cgi-bin/article.cgi?file=/chronicle/archive/2005/03/13/EDGSMAPC9G1.DTL.

13. "deals with abuses": Labaton, "Bankruptcy Bill Set."

14. "apparently, it is not": Robert Elder, "Judge Takes Congress to Task in Bankruptcy Case," Staesman.com, February 6, 2006, http://www.statesman.com/business/content/business/stories/other/02/5bankrupt.html.

15. "said the measure": Labaton, "Bankruptcy Bill Set."

16. "provide for no distinction" Jonathan Alter, "A Bankrupt Way to Do Business," Newsweek, April 25, 2005, http://www.msnbc.msn.com/id/7528519/site/newsweek/.

17. "Let's say Peter Jennings": Jonathan Alter, "A Bankrupt Way to Do Business," Newsweek, April 25, 2005, http://www.msnbc.msn.com/id/7528519/site/newsweek/.

18. "which have pushed": Kathleen Day, "Senate Passes Bill to Restrict Bankruptcy," Washington Post, March 11, 2005, http://www.washingtonpost.com/wp-dyn/articles/A24940-2005Mar 10.html.

19. gave $34 million more: Ibid.

20. 1 "the main lobbying": Labaton, "Bankruptcy Bill Set."

21. $150 million since 1990: Scheide, "You're Going Broke!"

22. Campaign Contributions to Senate Sponsor Chuck Grassley (R-IA): PoliticalMoneyLine from Congressional Quarterly, sampling of PACs for commercial banks and other lenders who contributed to Grassley, available at http://www.fecinfo.com/cgi-win/irs_ef_527 .exe?DoFn=&sYR=2004.

23. "Democrats Who Sold Out On Bankruptcy Protection And How Much They Got": U.S. Senate Roll Call Votes 109th Congress—1st Session, S 256 , March 10, 2005 as compiled through Senate LIS by the Senate bill clerk under the direction of the secretary of the Senate, available at http://www.senate.gov/legislative/LIS/roll_call_lists/roll_call_vote_cfm.cfm?congress=109&session=1&vote=00044; contribution information from the Center for Responsive Politics, specifically senator profiles listing commercial bank contributors at www.open secrets.org.

CHAPTER TEN

1. "Improving Federal Student": Office of the Press Secretary, "Fact Sheet: President Bush Signs the Deficit Reduction Act," February 8, 2006, available at http://www.whitehouse.gov/news/releases/2006/02/20060208-9.html.

2. Saving the Day: House Cuts Student Loan Rates: "Federal Family Education Loan Program (Ffelp)—Annual and Cumulative Commitments—Fy66-Fy2006," available at: http://www.ed.gov/finaid/prof/resources/data/06q4ffelpga.xls.

3. Forty-four percent: Kristin Davis, "Graduated Interest: Act Fast to Lock in College Loan Rates, Set to Rise on July 1," *Washington Post,* June 18, 2006.

4. Eighty-four percent: Elana Berkowitz and John Burton, "Burying Grads in Debt," Campus Progress of the Center for American Progress, November 28, 2005.

5. Two-thirds of all: Press release for the survey "The College Debt Crunch," "College Debt Has Profound Effect on Financial Security, Well-Being and Life Choices for Years to Come," AllianceBernstein, http://www.alliancebernstein.com/CmsObjectCDC/PDF/AB_CDC_Release.pdf.

6. "significant new benefits": Anne Marie Chaker, "Congress Cuts Funding for Student Loans," *College Journal* from the *Wall Street Journal,* December 23, 2005.

7. "students majoring in math": Lois E. Beckett, "House Republicans Pass Budget That Would Raise Student Loan Rates," *Harvard Crimson,* February 02, 2006.

8. $29,000 a year: Press release, "College Debt Has Profound Effect."

9. "College financing is": Ibid.

10. "paycheck to paycheck": Ibid.

11. Struggling Students: Ibid.

12. "federal government promises": Eileen Ambrose, "Time Is Now to Lock in Federal Student Loan Rates," *Chicago Tribune,* April 9, 2006 available at: http://www.orlandosentinel.com/business/yourmoney/sns-yourmoney-0409loans,0,169509.story?coll=orl-sns-yourmoney-headlines.

13. "Large lenders": Ibid.

14. "constitute an $86 billion": John F. Wasik, "College aid offices may not have best loan deals," *Bloomberg News,* July 31, 2006, available at http://www.bloomberg.com/apps/news?pid=newsarchive&sid=aOvPui4w5vto.

15. "There is a less": Leslie Stahl, "Sallie Mae's Success Too Costly? Does the Lender's Success Come at Too Steep a Cost to Students and Taxpayers?*" 60 Minutes*/CBSNews, May 7, 2006, available at http://www.cbsnews.com/stories/2006/05/05/60minutes/main1591583.shtml.

16. "Most universities guide": Ibid.

17. Sallie Mae manages: Ibid.

18. "All we were asking": Ibid.

19. "nearly a fifth": Ibid.

20. "Sallie Mae makes": Ibid.

21. $2.7 million to candidates: Ibid.

22. received over $200,000: Ibid.

23. "many school financial aid": Editorial, "Fortuitous Error: Congress Mistakenly Undercharged Some Student Loan Borrowers, Which Might Be Good News," *Washington Post,*

June 2, 2006, http://www.washingtonpost.com/wp-dyn/content/article/2006/06/01/AR 2006060101574.html.

24. "study after study": Ibid.

25. gave $52 million: Source: PoliticalMoneyLine from *Congressional Quarterly* at www.fec .info.com, specifically "Finance & Insurance Industry Totals" at http://www.fecinfo.com/ cgiwin/x_ee.exe?DoFn=06F.

26. "Donations to Congressional Candidates": Ibid.

27. Who Got the Money: "Member Money: Top Committee-Related Industries for Committee Members-Banking, Housing, And Urban Affairs, 109th Congress, 2006 Cycle Data," Center for Responsive Politics, http://www.opensecrets.org/cmteprofiles/profiles.asp?cycle=2006& CmteID=S06&Cmte=SBAN&CongNo=109&Chamber=S

28. $12.2 million from: Ibid.

CHAPTER ELEVEN

1. State Farm finally agreed: Kathy Chu, "State Farm Agrees to Pay Up for Katrina in Mississippi," *USA Today,* January 24, 2007, http://www.usatoday.com/money/industries/ insurance/2007-01-23-state-farm-katrina_x.htm?csp=34.

2. "Many homeowners have": Joseph B. Treaster, "Katrina: Hurricane Victims Say Agents Advised Against Flood Coverage," *New York Times,* July 14, 2006, available at: http://www.corp watch.org/article.php?id=13883.

3. "hurricane-force winds": Ibid.

4. more than $15 billion: Peter Henderson, "Homeowners Lose Katrina Insurance Food Case," Reuters, August 15, 2006, available at: http://planetark.com/dailynewsstory.cfm/newsid/ 37681/newsDate/16-Aug-2006/story.htm.

5. up to $30 billion: Treaster, "Katrina: Hurricane Victims Say."

6. "The fact of": Transcript, "Attorney Says Court Ruling Will Force Insurers to Pay Wind Damage, Despite Presence of Water Damage; Building Treehouses," CNN.com, August 19, 2006, http://transcripts.cnn.com/TRANSCRIPTS/0608/19/oh.01.html.

7. $130,000 of damage: Joseph B. Treaster, "Judge Rules for Insurers in Katrina," *New York Times,* August 16, 2006, http://www.nytimes.com/2006/08/16/business/16insure.html?ex= 1313380800&en=ee5920c47ca6894c&ei=5088&partner=rssnyt&emc=rss.

8. "Mr. Leonard, a police": Ibid.

9. "Nationwide and other": Joseph B. Treaster, "Katrina: Hurricane Victims Say Agents Advised Against Flood Coverage," *New York Times,* July 14, 2006, available at: http://www.corp-watch.org/article.php?id=13883.

10. fewer than 20 percent: Ibid.

11. "Judge Senter said": Treaster, "Judge Rules for Insurers."

12. "advisory in nature": Quincy Collins Smith, "Ruling Favors Insurer: But Judge Rejects Policy Language, Awards Money," *Sun Herald,* August 16, 2006, at http://www.sunherald.com /mld/sunherald/15283734.htm.

13. "The plaintiffs are": Henderson, "Homeowners Lose Katrina Insurance."

14. "talked him out": Treaster, "Katrina: Hurricane Victims Say."

15. "going to open": Treaster, "Judge Rules for Insurers."

16. "language in the": Joseph B. Treaster, "Judge Upholds Policyholders' Katrina Claims," *New York Times,* November 29, 2006, http://www.nytimes.com/2006/11/29/business/29insure.html?ex=1322456400&en=c20218c0a99b4c93&ei=5088&partner=rssnyt&emc=rss.

CHAPTER TWELVE

1. The number of adults: "Adult Cigarette Smoking in the United States: Current Estimates," CDC fact sheet, November 2006, http://www.cdc.gov/tobacco/factsheets/AdultCigarette Smoking_FactSheet.htm.
2. increasing the nicotine: Stephen Smith, "Cigarettes Pack More Nicotine," *Boston Globe,* August 30, 2006, http://www.boston.com/news/nation/articles/2006/08/30/cigarettes_pack_more_nicotine/.
3. "Most Smokers Use": "Research Report Series—Tobacco Addiction," National Institute on Drug Abuse, http://www.nida.nih.gov/ResearchReports/Nicotine/nicotine2.html#addictive.
4. Manipulating Nicotine to: Lois Keithly, PhD, et al., "Change in Nicotine Yields, 1998–2004," *Tobacco Control Program, Massachusetts Department of Public Health,* available at http://www.mass.gov/dph/mtcp/reports/nicotine_yields_1998_2004_report.pdf.
5. Massachusetts law requires: : Stephen Smith, "Cigarettes Pack More Nicotine," *Boston Globe,* August 30, 2006, http://www.boston.com/yourlife/health/other/articles/2006/08/30/cigarettes_pack_more_nicotine/?page=1.
6. The worst offender: Keithly, Ph.D, et al., "Change in Nicotine Yields."
7. "suppressed research": Henri E. Cauvin and Rob Stein, "Big Tobacco Lied to Public, Judge Says," *Washington Post,* August 18, 2006, available at http://www.truthout.org/cgi-bin/artman/exec/view.cgi/63/21910.
8. "selling a highly addictive": Philip Shenon, "New Limits Set over Marketing for Cigarettes," *New York Times,* August 18, 2006, http://www.nytimes.com/2006/08/18/washington/18tobacco.html?ei=5088&en=08c8bc3e2c0a9d51&ex=1313553600&adxnnl=1&partner=rssnyt&emc=rss&adxnnlx=1170449193-3AhOyrpWYEjD1hozDm5sWg.
9. "Defendants purposefully designed": *United States of America v. Philip Morris et al.,* Civil Action 99-2496 (GK), Final Opinion, United States District Court for the District *of Columbia,* 351, available at http://www.tobaccofreekids.org/reports/doj/FinalOpinion.pdf.
10. "In short, defendants": Ibid, 4.

CHAPTER THIRTEEN

1. routinely accepted bribes: Correspondent, "Cheques, lies and videotape," BBC News, September 28, 2000, http://news.bbc.co.uk/1/hi/world/europe/946938.stm.
2. Schröder announced his: Craig Whitlock and Peter Finn, "Schröder Accepts Russian Pipeline Job: German Critics Charge Conflict," *Washington Post,* December 10, 2005, http://www.washingtonpost.com/wp-dyn/content/article/2005/12/09/AR2005120901755.html.
3. a $310,000 salary: "Germans want probe of 'Schrodergate'," United Press International, April 26, 2006, available at http://freerepublic.com/focus/f-news/1622761/posts.
4. from Portovaya Bay: Simon Araloff, "Schröder-Putin Pact: Germany and Russia Divide

Europe Again," Axis Information and Analysis, April 5, 2005, http://www.axisglobe.com/article.asp?article=51.

5. 200 kilometers long: "Northern European Gas Pipeline (NEGP)," http://negp.info/news/news1.html.

6. "it is expected": Ibid.

7. "The only possible": Editorial, "Gerhard Schröder's Sellout," *Washington Post,* December 13, 2005, http://www.washingtonpost.com/wp-dyn/content/article/2005/12/12/AR20051 21201060.html.

8. "The North European": "Northern European Gas Pipeline (NEGP)."

9. "Many have wondered": Editorial, "Gerhard Schröder's Sellout."

10. "It stinks": "Schröder Attacked over Gas Post," BBC News, December 10, 2005, http://news.bbc.co.uk/2/hi/europe/4515914.stm.

11. "position is not": Craig Whitlock and Peter Finn, "Schröder Accepts Russian Pipeline Job," *Washington Post,* December 10, 2005, http://www.washingtonpost.com/wp-dyn/content/article/2005/12/09/AR2005120901755.html.

12. "Schröder was such": Ibid.

13. "Schröder and the": Ibid.

14. officer in the Stasi: "Dresdner Bank Head Warning Linked to Putin's Spy Past—Paper," MOSNews.com, February 24, 2005, http://www.mosnews.com/news/2005/02/24/stasi.shtml.

15. "Germany defend its": "EU Leans on Germany over Russian Loan," United Press International, May 10, 2006, available at http://www.redorbit.com/news/science/498486/eu_leans_on_germany_over_russian_loan/index.html.

16. "Neelie Kroes": Ibid.

17. "large state aid": Ibid.

18. "The commission requested": Ibid.

19. "Referenda have not": [Name?], interview with Dick Morris, April 2003.

20. "Our task, as": Ibid.

21. "incompatible with his": Encyclopedia excerpt, "Jacques Chirac," *Wikipedia,* http://en.wikipedia.org/wiki/Jacques_Chirac.

22. "empty chairs": Colin Randall, "President's Men Tumble in Chirac Sleaze Trial," *The Telegraph,* http://www.telegraph.co.uk/news/main.jhtml?xml=/news/2005/10/27/wfran27.xml.

23. Chirac's Partners in Crime: Ibid.

24. "dilapidated sixteenth century": Jon Henley, "How to Succeed in Politics Without Really Lying: The charmed Career of Jacques Chirac," FrontPageMag.com, April 19, 2002, http://www.frontpagemag.com/GoPostal/commentdetail.asp?ID=6175&commentID=61971.

25. "in case something": Correspondent, "Cheques, lies and videotape," *BBC News,* September 28, 2000, http://news.bbc.co.uk/1/hi/world/europe/946938.stm.

26. "in Chirac's presence": Encyclopedia excerpt, "Corruption scandals in the Paris region," *Wikipedia,* http://en.wikipedia.org/wiki/Corruption_scandals_in_the_Paris_region.

27. "judge Halphen's hands": Ibid.

28. "I have proof": Henley, "How to Succeed in Politics."

29. "personally informed of": Ibid.

30. "the inspiration": Ibid.

31. "lies, calumny, and manipulations": Ibid.
32. "Me, eat jam?": Ibid.
33. Mery turned the tape: Correspondent, "Cheques, Lies and Videotape."
34. The tax bill was: Ibid.
35. Strauss-Kahn was: Ibid.
36. "either through legal": Encyclopedia excerpt, "Corruption Scandals in the Paris Region," *Wikipedia.*
37. the RPR got 1.2 percent: "Chirac Allies in Corruption Trial," BBC News, March 21, 2005, http://news.bbc.co.uk/2/hi/europe/4367199.stm.
38. prominent French companies: "Corruption Scandals in the Paris Region."
39. "towered" over the: Ibid.
40. Charles Pasqua was: Joseph Fitchett, "Exiled Fugitive Threatens to Tell All in Chirac Case," *International Herald Tribune,* February 5, 2002, at http://www.iht.com/articles/2002/02/05/paris_ed3_.php.
41. "I have never": Nicholas Le Quesne, "Jacques Goes on the Attack," *Time Europe,* February 26, 2002, http://www.time.com/time/magazine/article/0,9171,901020225-213187,00.html.
42. "by fabricating evidence": Fitchett, "Exiled Fugitive Threatens."
43. "core allegation depicts": Ibid.
44. "that during his": Henley, "How to Succeed in Politics."
45. Those prosecuted included: "Juppe Faces Suspended Sentence," BBC News, October 10, 2003, http://news.bbc.co.uk/2/hi/europe/3182226.stm.
46. "spelling and grammatical": "Corruption scandals in the Paris Regions."
47. Parisian mayor Bertrand Delanoe: Ibid.
48. "buy airline tickets": Suzanne Daley, "Heat Is on French President: Plane Tickets for Chirac's Family Called into Question," *San Francisco Chronicle,* June 28, 2001, http://www.sfgate.com/cgi-bin/article.cgi?file=/c/a/2001/06/28/MN113896.DTL.
49. "see how cash": Ibid.
50. "fraud and favoritism": Alexandra Fouché, "Q & A: Chirac's Corruption Battle," BBC News, October 10, 2001, http://news.bbc.co.uk/1/hi/world/europe/1448471.stm.
51. "I no longer": Greg Frost, "Tales of Sleaze Cloud French Elections," Reuters, March 3, 2002.
52. "Under France's constitution": Fouché, "Q & A: Chirac's Corruption."
53. "However, the constitution": Ibid.
54. The Constitutional Council: Alexandra Fouché, "Alexandra Fouché looks at what that means for Mr Chirac and French politics ahead of this year's elections," BBC News Online, October 10, 2001.
55. detail of history": "Backgrounder: Jean-Marie Le Pen and the National Front," Anti-Defamation League, April 23, 2002, http://www.adl.org/international/le-pen_new.asp.

CHAPTER FOURTEEN

1. "our consumers' addiction": Transcript, *Lou Dobbs Tonight,* CNN.com, June 22, 2006, http://edition.cnn.com/TRANSCRIPTS/0606/22/ldt.01.html.
2. "Over the past five": W. Michael Cox and Richard Alm, "The Fruits of Free Trade," Federal Reserve Bank of Dallas, 2002, http://www.dallasfed.org/fed/annual/2002/ar02.pdf.

3. Take sugar as an example: Ibid.

4. "it would cost": Ibid.

5. frozen concentrated orange juice: Ibid.

6. The Cost of Protection: Ibid.

7. "subsidies to steel-producing": Ibid.

8. sugar political action committees: "Sugar: Long-Term Contribution Trends at http://www.opensecrets.org/industries/indus.asp?Ind=A1200 and "Sugar: Top Contributors to Federal Candidates and Parties," http://www.opensecrets.org/industries/contrib.asp?Ind=A1200&Cycle=2004, Center for Responsive Politics.

9. "a small number": Cox and Alm, "The Fruits of Free Trade."

10. the cost of sugar: Elizabeth Querna, "No, It's Not Just Your Sweet Tooth," *U.S. News & World Report,* March 28, 2005, http://www.usnews.com/usnews/health/articles/050328/28sugar.htm.

11. "When I got on": Hedrick Smith, "Sugar Daddy," PBS.org, March 25, 1997, http://www.pbs.org/newshour/bb/congress/march97/sugar_3-25.html.

12. gave $2.3 million: "PAC Hard Dollar Contributions Made by Industry Groupings (Agriculture)," PoliticalMoneyLine from *Congressional Quarterly* at FECinfo.com, specifically at http://www.fecinfo.com/cgi-win/x_ee.exe?DoFn=06A.

13. timber PACs gave $831,000: "PAC Hard Dollar Contributions Made By Industry Groupings," PoliticalMoneyLine from *Congressional Quarterly* at FECinfo.com, specifically at http://www.fecinfo.com/cgi-win/x_sic.exe?DoFn=.

14. $303,000 to protect: Ibid.

15. trade protectionism costs: Cox and Alm, "The Fruits of Free Trade."

16. almost 50 million jobs: Ibid.

17. Farm subsidies in Europe: Joseph E. Stiglitz and Andrew Charlton, *Fair Trade for All: How Trade Can Promote Development,* (New York: Oxford University Press, 2006), 218.

18. "From 1995 to 2002": Nicolas Heidorn, "The Enduring Political Illusion of Farm Subsidies," *San Francisco Chronicle,* August 18, 2004, available at http://www.independent.org/newsroom/article.asp?id=1340.

19. "these subsidy programs": Brian Riedl, "Still at the Federal Trough: Farm Subsidies for the Rich and Famous Shattered Records in 2001," Heritage Foundation, April 30, 2002, http://www.heritage.org/Research/Agriculture/BG1542.cfm.

20. $24 billion a year: Jonah Goldberg, "Welfare Kings: Boy, Do They Spread Themselves Out," National Review Online, August 4, 2006, http://article.nationalreview.com/?q=NWI3ZTMyZDYyMTRmOWZmYTM4ZGIwNGQ2YTYzNzU0NmE=.

21. Sixty percent of farms: Bill D'Agostino, "Farms in Palo Alto?" *Palo Alto Weekly* Online, March 09, 2005, http://www.paloaltoonline.com/weekly/morgue/2005/2005_03_09.farm09mb.shtml.

22. more than a million: Heidorn, "The Enduring Political Illusion."

23. twenty-five congressional districts: "Uncle Sam's Teat: Can America's Farmers be Weaned from Their Government Money?" *The Economist,* September 7, 2006, http://www.economist.com/world/na/displaystory.cfm?story_id=7887994.

24. This chart shows: Ibid.

25. "large farms and agribusinesses": Riedl, "Still at the Federal Trough."

26. Subsidies They Don't Deserve: Ibid.

27. more than $4 million: Andrew Backover, "Ebbers Linked to over $4M in Farm Subsidies," *USA Today*, November 12, 2002, http://www.usatoday.com/money/industries/telecom/2002-11-11-ebbers_x.htm.
28. "Since 2000, the U.S.": Goldberg, "Welfare Kings: Boy, Do They Spread Themselves Out."
29. only 60 percent produced: Cox and Alm, "The Fruits of Free Trade."
30. a quarter of our cars: Ibid.

INDEX

Mynderse Academy HS Library

SFS0023099

MYNDERSE ACADEMY HS LIBRARY
105 TROY STREET
SENECA FALLS NY 13148